Scribe Publications
MATESHIP WITH BIRDS

A. H. (ALEXANDER HUGH) CHISHOLM was born in Maryborough, Victoria in 1890, and worked on the *Maryborough Advertiser* before moving to Brisbane to work on the *Daily Mail*, and subsequently to Melbourne to edit the *Argus*. Chisholm worked with C. J. Dennis and published his major work, *The Making of a Sentimental Bloke*, in 1946. Chisholm died in 1977.

SEAN DOOLEY is a Melbourne comedy writer and author whose first book, *The Big Twitch*, outlined his attempt to break the Australian birdwatching record. Sean is currently editor of *Australian Birdlife*, the magazine for BirdLife Australia.

The Author at a Shrike-tits' Nest.
(Photo. by N. M. Chisholm.)

Mateship *with* Birds

A. H. Chisholm

SCRIBE
Melbourne • London

Scribe Publications Pty Ltd
18–20 Edward St, Brunswick, Victoria, Australia 3056
Email: info@scribepub.com.au

First published by Whitcombe and Tombs Limited 1922

Published by Scribe 2013

Text copyright © Alec Chisholm 1922
Text design copyright © Scribe Publications 2013
Foreword copyright © Sean Dooley 2013

All rights reserved. Without limiting the rights under copyright
reserved above, no part of this publication may be reproduced,
stored in or introduced into a retrieval system, or transmitted,
in any form or by any means (electronic, mechanical, photocopying,
recording or otherwise) without the prior written permission
of the publishers of this book.

While every care has been taken to trace and acknowledge copyright,
we tender apologies for any accidental infringement where copyright
has proved untraceable and we welcome information that would redress
the situation.

Typeset in 12/14.5 pt Bell MT Std by the publishers
Printed and bound in China by 1010 Printing

National Library of Australia
Cataloguing-in-Publication data

Chisholm, Alec H. (Alec Hugh), 1890-1977.

Mateship with Birds / A. H. Chisholm.

9781922070326 (hbk.)

1. Birds–Australia.

598.0994

Other Authors/Contributors: C. J. Dennis; Sean Dooley.

www.scribepublications.com.au

CONTENTS

List of Illustrations   vii

Foreword by Sean Dooley   ix
Introduction by C. J. Dennis   xv
Preface   xvii

PART I. — A PAGEANT OF SPRING

I. The Gifts of August   3
II. September Revelry   17
III. October the Witching   31
IV. The Passing   43
V. With Children in Birdland   57

PART II. — BIOGRAPHIES OF BIRDLAND

VI. The Idyll of the Blossom-Birds   73
VII. The Aristocracy of the Crest   87
VIII. Days among the Robins   105
IX. Fine Feathers and Fine Birds   121
X. The Spirit of Australia   137
XI. The Paradise Parrot Tragedy   153

Index   169
List of Scientific Names   176

TO THE MOTHER AT HOME

LIST OF ILLUSTRATIONS

Frontispiece: The Author at a Shrike-tits' Nest

Between pages 42 and 43

Jacky Winter's Rostrum
Jacky Winter at Home
"The madcap New Holland Honeyeater"
Honeyeaters' Home amid Wax-flowers
Heath-gatherers on the Mountains
A Guest of the Heath-bells
Yellow Robin brooding
"Double-decked" Nest of Tit-warbler
A Lesson in the Bush
A September Outing
Nest of Rufous Song-lark
Nest of Babbling-thrush
Nest of Grey Thrush
"Georgie," the Reed-warbler
Wagtail on nest in flowering apple-tree
White-shouldered Caterpillar-eater
English Blackbirds' Nest in eucalypt
Wood-swallows' Nest
Wood-swallow at Home
Wood-swallows' Nest in an Orchard
Caterpillar-eater at Nest
Female Caterpillar-eater on Nest
Tree-creepers' Home
Nest (opened out) of Spotted Diamond-birds
Baby Crested Bell-bird
Baby Yellow Robins
Nest of Red-capped Robin
Female Red-breasted Robin brooding
Kookaburra and young
Hungry baby Red-caps

Between pages 106 and 107

Mateship with Mother Robin
"Talking birds" on the edge of a Queensland jungle
Calling on the Grey Thrush
Calling on the Tit-warbler
The Spotted Bower-birds' visitor
Marjorie and mother Dove
"See!" At a Honeybirds' home
White-plumed Honeybirds' Nest in a dry bush
The dainty Sun-bird at home
Yellow-tufted Honeybirds' Nest
White-plumed Honeybirds' Nest
Fuscous Honeybirds' Nest
Happy Australians. Domesticated Cockatoo
Baby Spine-tailed Logrunner
The jungle home of the Whip-bird
Edge of the jungle. Where Whip-birds and Shrike-tits meet
A rare quartette
Shrike-tits' nest
Nest of the Crested Bell-bird
The famous Whip-bird
Shrike-tit and young
Mother-love. Yellow Robin gazing down babe's mouth
Yellow Robin admiring its eggs
Yellow Robin "posing"
Pale-yellow Robin on nest in jungle
Yellow Robin feeding mate
Watching Rose Robins
Where Golden-breasted Whistlers abound
Female Golden-breasted Whistler
Gilbert Whistlers' nest
Rufous Whistler at nest
Rufous Whistler's mate
Photographing a Rufous Whistlers' nest
Magpie-lark and family
The Magpies' Banquet
Pied Butcher-bird
Restless Flycatcher
Wagtail brooding
Wagtail protecting brood from sun
White Watsonias on a Queensland Mountain
Habitat of Paradise Parrot
Paradise Parrots

FOREWORD

I KNEW ALEC Chisholm's name before I knew the name of any other birdwatcher — indeed before I knew the names of many birds. When I was eleven, my parents bought me my first 'real' bird book, the 1976 edition of the *Reader's Digest Complete Book of Australian Birds*. In it was a foreword by Alec Chisholm. He was mentioned again in the account of the Paradise Parrot as the last experienced birder to have seen this most beautiful bird before it became extinct (a fact that the dogmatic Chisholm refused to acknowledge right up until his death more than fifty years after his 1922 sighting). That alone fired my budding birdwatcher's brain with awe.

It is only in recent years, however, as I have delved into the history of ornithology in Australia as the author of my own books on birdwatching, and as the editor of *Australian Birdlife*, that I have come to appreciate just what a colossal figure the man referred to as "Chis" was — right up there with John Gould or Neville Cayley in terms of importance to Australian bird study and for providing a voice for Australian birds.

That Chisholm ever came to be held in such esteem is quite remarkable. In an era when the professional ornithologist ruled the roost, Alec Chisholm had no scientific training. It was in his writings, both for ornithological journals and in his innumerable newspaper articles and many books, that Chisholm left an indelible mark.

Chisholm was never a likely candidate to become such a hugely popular and influential writer who counted amongst his circle literary luminaries such as C. J. Dennis (who penned the foreword to the initial publication of *Mateship with Birds* in 1922) and Dame Mary Gilmore. Born in the central Victorian goldfields town of Maryborough, Chisholm left school at the age of twelve. After stints as a delivery boy and coach-builder's apprentice, he almost fell into journalism through his passion for birds.

While his observations of birds encountered in his bush wanderings first started to appear at a precociously young age in journals such as *The Emu*, it was the publication of his first story in a general newspaper — a passionate exhortation to stop the slaughter of egrets for plumes used for decoration in the millinery trade — that set his career as a journalist rolling. It was to be a long and distinguished career as a reporter and editor for newspapers in three states, including stints as a sports reporter in Melbourne.

Chisholm went on to be the press officer for the governor-general, the editor of *Who's Who in Australia*, and the editor-in-chief of the *Australian Encyclopaedia*. But it is for his nature writings, particularly about birds, that Alec Chisholm's legacy endures. In almost 70 years of publication, he touched and inspired generations of Australians, sharing his boundless passion for birds. While his accounts may at times seem a trifle florid to the contemporary reader, they were a welcome relief to the stuffy formality and borderline pomposity of many ornithological writings of the day. One can see that, in his writing style, Chisholm was searching for a language as rich as the birdsong he so loved.

And the range of his references, typical of such a voracious autodidact, is both egalitarian and eclectic.

FOREWORD

Chisholm happily allows snippets from Wordsworth, Shakespeare, and Greek mythology to merge with observations taken from serious ornithological literature, quotes from local schoolchildren, and examples of Indigenous bird knowledge to instil in the reader the sense of wonder that Australian birds could evoke.

Eschewing the pomposity of scientific nomenclature, Chisholm's style was direct and passionate, conveyed in a plain language that could be understood by all readers. The simple title, *Mateship with Birds*, speaks volumes — unlike the overly descriptive titles given to most bird tomes of the era, this is unadorned and direct, focusing on the connection between people and birds through the concept of mateship with which Australians so often identify.

As an example of his popular touch, thornbills are referred to throughout this book as tit-warblers — a name that Chisholm knew would register with the majority of his readers — even though he sat on a committee that, given the job of codifying vernacular names for Australian birds, had designated the name 'thornbill' for the indigenous family of small, insectivorous birds.

His commitment to the popularising of bird study did not always make him popular with the professional ornithologist set. The perceived stereotype of the birdwatcher is of a gentle soul, communing with nature. In the internecine world of birdwatching politics, however, nothing could be further from the truth. Chisholm was as willing a participant in a good stoush as any, being frequently described as querulous and quarrelsome. Others noted his feisty and obdurate nature, and he was certainly tenacious in his pursuit of what he felt was the correct course of action — such as his vehement opposition to egg collecting and what he regarded as the unnecessary shooting of birds for 'scientific' specimens.

Such clashes led to insinuations that Chisholm was pandering to populist sentiment, but what still comes through powerfully in his writing is that he was a bridge between the arcane academic world of ornithology and the general public. He seemed to innately understand that all the knowledge in the world is useless if cloistered among a select few.

To this end, Chisholm was always a strong advocate for inspiring in our children a love of, and respect for, our birds. In our current age of "nature deficit disorder", when children are more than ever cut off from direct experience of the physical world, works such as *Mateship with Birds* are needed more than ever to spark the embers of interest in the natural world.

With his direct experience of the Paradise Parrot slipping away into extinction — a subject first aired in this book, and one that he would return to throughout the rest of his life — Chisholm was one of the first to make an impassioned plea to save our unique wildlife. What makes for poignant reading 90 years on is that many of the birds that are centre stage in *Mateship with Birds* are now in decline. For example, the Regent Honeyeater's nesting habits are almost casually mentioned, with the author assuming a certain familiarity with the species amongst his readers; but today, this stunning woodland bird is down to around 500 individuals in the wild, and despite great efforts by many volunteers and conservation bodies such as BirdLife Australia, appears to be on a downward spiral towards extinction. The Crested Bellbird (the one that adorns the sides of its nest with live caterpillars) has dropped out of many of the landscapes with which

FOREWORD

Chisholm was familiar. And the first bird mentioned in Chapter 1 — the Jacky Winter — no longer sings its 'ecstatic' spring song in many of Chisholm's former haunts. Were he to return today, Alec Chisholm would indeed find the spring far more silent than he ever could have imagined.

It is with gratitude that I greet the re-publishing of this book, which itself has lain silent for too long, and I hope that in discovering it, the contemporary reader is infected with Alec Chisholm's enthusiasm for birds in the way that our grandparents were. For unless we start to care, it is frightening to contemplate how many more spring voices will be lost to future generations.

Sean Dooley
December 2012

INTRODUCTION

It is many years since I first heard of the fame of Mr. A. H. Chisholm, who has honored me with a request to write a preface for his book on birds. I say "honored" because he might easily have found a man capable of writing a worthier preface than mine to such a worthy book. I value the compliment the more, because it was mainly through the writings of Mr. Chisholm that I began to take an interest in Australian birds, discovering a new pleasure, and one that will always endure.

Many books have now been written about Australian birds. I suppose nearly all these books are chiefly of interest to the scientist, who delights in naming a live bird in a dead language. But, by the layman amongst bird lovers — whose interest is no less keen because non-scientific — such a book as *Mateship with Birds* will be most heartily welcomed.

Although, in regard to the scientific aspect, Mr. Chisholm is generally regarded as one of the foremost authorities upon birds in Australia, he is not content with mere classification and tabulation. This book of his has been, in a sense, an inspiration. Not only will it be of great interest to the bird lover and the bird observer, but many a man who hitherto has taken little interest in such things will find here new interests and fresh delights. Even he, whose only acquaintance with birds has been amongst sparrows of the city streets, and an occasional pigeon or blackbird, will discover unexpected entertainment.

The title is a particularly happy one, for it indicates the author's fraternal attitude and methods. Many a learned savant shoots birds with a gun and writes about them as a pedant. Mr. Chisholm shoots them with a camera and writes about them as a human being.

This title reminds me of a certain good mate of mine — a grey thrush who came to me regularly each morning for his breakfast. He ate confidently from my hand, and, having eaten, piped a song of thanksgiving. His name was "George." A prowling cat got him in the end. A book upon "Mateship with Cats" would earn my hearty disapproval.

It is a good thing that Australians, during recent years, have taken a vastly increased interest both in the flora and in the birds of their native land. So far as the birds are concerned, this book will do much to stimulate that interest. Here is a human story, and an entertaining story, written by one who has an accurate knowledge of his subject.

For these reasons, and for the sake of our mates the birds, it is recommended not alone to bird lovers, but to every Australian who has taken an interest in birds. And how many amongst us have listened to the mellow fluting of the grey thrush at eve, or the wild, free piping of magpies at morn, and have remained unmoved?

C. J. Dennis
(*"The Sentimental Bloke"*)
Melbourne

PREFACE

Rather many (but not excessively many) years ago, ere the facts and fancies embodied in this book had fairly begun to take form, I read in a preface to a work on natural history the statement that much of the matter contained in the chapters had been written in the open air. That appealed as a pleasant boast and a worthy example. How fine it would be, I thought, to produce a book, be it ever so humble, that had been written "under the splendid sky"!

But the years between brought a change of attitude. They also brought a change of circumstances. There is no hesitation in confessing, now, that practically the whole of this book has been written, subject to the demands of daily journalism, in the scarcely hallowed atmosphere of a city boarding-house!

The reader may be inclined to think that the change of attitude should be cited as a corollary, and not as a prelude, to the change from bush to city life. Really, though, the altered situation had little to do with the case. The flowers that bloom in the Spring were much more responsible. The birds that nest in the Spring were the deciding factor.

Is there any genuine bird-lover who could sit and write, more or less stolidly, while here, there, and everywhere the play of Nature was in active progress? To attempt to do so were rather like discussing Puck while Ariel held the stage. At all events, practically the only writing I have done in the bush (that is, while among the wild birds and flowers), has been to jot down odd notes — fleeting impressions and details that might be lost in a maze of incident and minor adventure before there was time to "post up" the regular note-book.

Yet I would not have it thought that the pages of this book reflect electric light rather than the sun. For the most part, at least, the knowledge here presented was never gained of schools nor books; and it is given forth in the hope that sufficient of the freshness of bush mornings remains to counteract any drabness that may have accumulated from the "ripening" influence of city and study. True enough, many of the chapters are "bookish" in the sense that verse quotations are numerous. But for this the poets are as much to blame as were the birds and the flowers in another aspect. Blessings upon all three for Springtime (and world-wide) monopolists of youthful heads and hearts!

It was a custom of mine in the frankly impressionable period — the years immediately succeeding the catapult stage — to go bushwards with camera and field glasses over shoulders and a book of good verse or prose in a pocket. Leigh Hunt recommended much the same treatment for Shakespeare on Shakespeare's birthday, but held himself ready to drop the book at the call of living Nature. Similarly, my reading in the Australian bush has been entirely desultory. Did the birds prove unusually coy, the sunlight would play on the pages of the book for quite a long time. But there were occasions when, in a manner of speaking, the curtain rose a few minutes after the spectator was seated. Reading was usually out of the question then. Probably the book had been read before and would be read again, but it did not follow that the precise little secrets of the bush being revealed would ever be chanced upon again. Withal, a good deal of reading was accomplished in all those odd moments, and so the echoes from the poets are portions of the gleanings of early bush days. The affinity seems natural enough, too; certainly, one could say of lyric verse, equally with the songs of birds, "The music in my heart I bore long after it was heard no more."

It is quite true, as the late John Burroughs remarked long ago, that you cannot run and read the book of Nature. Too many would-be naturalists, lured by the greenness of distant fields, rush about their own and other countries, to the neglect of the opportunity for more fraternal study near at hand. I have sought acquaintance with wild birds in many parts of Australia, but intimacy has come only by dalliance. This is the kind of thing, of course, that old Belarius, of Shakespeare's *Cymbeline*, preached to the two young princes whom he had kept in the forest since babyhood, causing them to reply, bitterly:

> Out of your proof you speak; we, poor unfledged,
> Have never wing'd from view o' the nest. . . .

But I am far from decrying the advantages of travel to the naturalist. The point is that close knowledge of particular birds comes from concentration, and that a bush dweller who applies himself to his own locality can learn things — excellent things — that are denied men who traffic up and down and, maybe, "translate" Nature by means of mere lists and catalogues. How seldom this is conceded by the stay-at-home, much less by the traveller!

Some years ago, when attempting to get together the life-history of a certain bird, I wrote a naturalist who had studied the species, for corroboration on one or two points. The reply was a little startling. My friend had given up bird-study for the time being, it appeared, because he had "worked out" his district. Save us from such shallow beliefs! There is no such thing as working out the ornithological interest of a district. It is possible, by constant watchfulness, to achieve something like a complete record of the birds in a certain area, but the character and habits of those birds are studies that age cannot wither nor custom stale.

Sir Wm. Beach Thomas, who is at present in Australia, has given me an eloquent instance on the point. He wrote a paragraph for the London *Daily Mail* dealing with the song of the Blackbird, and within a day or two he received 160 letters on the subject — a host of varying opinions. There you have a bird-voice among the most familiar in the world, something as old as British history, but one that is not yet understood, not yet humanised, not yet "captured." Who shall say, then, that the birds of any district in this young land are "worked out" on mere bowing acquaintance?

One of our leading ornithologists observed recently that Australian birds whose complete life histories had been written could be counted on the fingers of one hand. That is true enough, and it brings us again to the need for concentration in study — to the fact that there is usually more intimate natural history to be gained from an hour's repose in a particular spot than in a whole morning of hurry. No right-thinking bird has a lasting dislike for the human figure; it is the noise and movement in association that arouse fear and resentment. Nor is the enjoyment of a bush ramble lessened by the cultivation of quietness. More than one troop of hearty boys have given

PREFACE

me concrete evidence on the point; I have even known it to be achieved by a bevy of feminine school-teachers! But the bird observer certainly has more scope for study when alone. It is thus that most of the material contained in this book has been gathered.

No such claim is made for the story of the Paradise Parrot, of Queensland. Intervals of some years were devoted to the strange, eventful history of this bird, but the chapter itself indicates that information was not gained at first-hand. This story was included, at the last minute, in deference to the wishes of friends who desired to have it in more permanent form than that provided by a magazine.

Of the other sketches, several appeared in slightly different form in the Sydney *Mail*, one in *Everylady's Journal* (Melbourne), and certain more technical portions in *The Emu*, the quarterly journal of the Royal Australasian Ornithologists' Union. The permission to reprint is in keeping with the fine service these publications have rendered to the cause of nature study in Australia. I am also indebted to my friends, Messrs. D. W. Gaukrodger, C. H. H. Jerrard, W. G. and R. C. Harvey, and J. H. Foster (Queensland), R. T. Littlejohns, S. A. Lawrence and L. G. Chandler (Victoria), and Sid. W. Jackson (New South Wales), for the photographs which bear their names.

Brisbane, 1922
A.H.C.

PART I

A PAGEANT OF SPRING

Chapter I

THE GIFTS OF AUGUST

I like to think that O. Henry was not altogether facetious in laying it down that the true harbinger of Spring is the heart. "It's just a kind of feeling," he confides. . . . "It belongs to the world." At any rate, one may nod sagely to these observations without necessarily subscribing to a further suggestion, that the three kinds of people who feel the approach of Spring first are poets, lovers, and poor widows — which is another question, one quite beyond me.

It is clear enough, however, that the manner of advent, as distinct from the appeal, of the sweet season is not the same in varying latitudes. Spring returns to Southern Australia, for instance, with grace rather than might. There is not the Swinburnian clamor — the "noise of winds and many rivers" — which marks the breaking of Winter's sway in the old world, and, on the other hand, the semi-languid nature of the Spring of Northern Australia is pleasantly lacking. But in the southlands of this great Commonwealth at least there is good red-blooded vitality, almost sufficient of inspiring throb to make the approach of Persephone perceptible to the dullard.

> 'Twas Jack o' Winter hailed it first,
>   But now more timid angels sing:
> For what dull ear can fail to hear
>   Afar the fluting of the Spring?

All the Winter through Jacky of that ilk, the Brown Flycatcher, has been disporting himself, like Lowell's Blue-Bird, "from post to post along the cheerless fence," or chanting his "Jacky-Jacky-Jacky, Sweeter-sweeter-sweeter!" in the big dry tree by the roadside. Now his penetrating pipe has left the philosophic key and risen close to the ecstatic, striking a responsive chord in the breasts of those gems of the grass, the communistic Red Robins, White-fronted Bush-Chats, and Yellow-tailed Tit-Warblers.

All the Winter through, too, the merry-making Honey-Birds have been playing and chortling about the blossoming eucalypts. Is it merely fancy that on these crisp, bright mornings of early August their notes have taken a higher range, or are they, in truth, sensitive of the fact that nesting-time is near? Possibly it is because the Honey-Birds live "closer" to the sun that they, rather than other birds of the Winter, are quick to detect the approach of the nearing Spring. But the insectivorous birds of the ground have noted a quickening of the pulse of the earth, reflected in the steady increase of insect-life, and, perhaps, the opening of the cheerful-looking flowers of the sundew and "early Nancy." And none is more apt to spread the glad tidings than the Brown Flycatcher. The fact that Jack o' Winter is much more quietly garbed than the other small species matters least of all to him; he is

THE GIFTS OF AUGUST

just as companionable as any of them, and a good deal more assertive.

I remember being much entertained in a far-off August by the antics of one of these little brown sprites, which insisted in dancing attendance on a distant relative, a Restless Flycatcher to wit. Jacky's motives may have been purely disinterested, but it seemed to me he cherished a shrewd idea that the curious rasping notes or whirring wings of the "Scissors-Grinder" would disturb sufficient insects for them both. Some element of the same suspicion evidently possessed the black-and-white bird. He showed no appreciation whatever of his relative's company, and once or twice he snapped viciously at the uninvited guest who sat so fraternally near. But little amenities of that kind left the philosophic Jacky quite unruffled. A veritable Elisha in point of determination, he was flitting serenely along after his ungracious companion when I lost sight of both.

A curiously distant bird altogether is the Restless Flycatcher. Unlike its double, the familiar Willy-Wagtail, it suggests but little, if any, of the *human* quality one subconsciously looks for in the voice of a bird, and it is not easy to get on "speaking" terms with either the spanking male or buff-tinged female. Only once have I succeeded in calling a "Scissors-Grinder" into conclave, and then it was purely curiosity at the gathering of other birds that persuaded the lady to stay awhile. On one other occasion, however, a male bird unwittingly gave me a close audience at its "wheezing" performance, and I saw how the strange notes worked up from the body of the bird until, at their height, the small bill was wide open, as with a brooding bird gasping in the heat of the Summer sun. It is not true, as several books assert, that the grinding notes which have made the Restless Flycatcher famous are only uttered when the bird is hovering a few feet above the ground. I hear them just as often when their author is perched on stump or fence-post. Essentially, however, this beautiful Flycatcher has the sea-birds' manner of hunting; it is able to look down in flight without turning the head to one side.

Soon, now, the beauteous Robins will be unobtrusively moving off to the uplands for the breeding-season; the Flycatchers, too, will be seeking out their old nesting-trees, and the more open areas will be left to the busy Magpies, the shapely Bush-Chats, and the merry Tit-Warblers. What more sparkling bird-melody than the voices of the Yellow-tailed Tit-Warblers giving greeting to our Lady of the Spring! You hear them now in almost every timbered field and along the skirts of any road or highway bounded with hedge-accommodation — happy bursts of melody akin to the softly joyous laughter of young girls.

It is as yet too early for the migratory and travelled birds generally to reappear. If the Winter be unusually mild, it is possible that an enterprising Lark or Warbler will be heard some weeks in advance of the main body of those birds; but the only migrants regularly presented by August to the maiden Spring in Central Victoria are the Cuckoos. So soon as the impalpable heralds of the coming season begin to throb in the frosty air, the bird-lover finds himself straining an attuned ear for the call of the Pallid Cuckoo. He has not long to wait. Be it wet or fine, the airy voice of this Ishmael of the bird-world will certainly be heard before the month is out, and its cumbersome form — "floppy," a critical girl has called it — with its attendant bevy of resentful smaller species, will be seen bunched upon fence, telegraph line, or any other point of vantage in the sunlit woods.

I can find nought of intrinsic melody in this Cuckoo-voice, or, indeed, in the call of any of its close relatives. But the leisurely, throaty strain, if it has not the blitheness that poets have found in the wandering voice of the famous bird of the old world, has a persuasive quality of its own at this time o' year, something that speaks — without, shall we say? attempting to do so — of pulsing life in sun-warmed fields. The Winter of 1909 was exceptionally wet in Victoria, but not sufficiently so to repress this herald of the Spring. On the twelfth day of August there rang out, clear and high above a factory which endured my services, the invitation of a Pallid Cuckoo. No shades of a prison-house closed about *one* growing boy that afternoon; he was out along a bush railroad superintending a disagreement between a philosophic grey Cuckoo and a host of excited Honeyeaters!

The smaller Bronze Cuckoo I have found to be less definite in its time of arrival, my earliest record for Central Victoria being the eighth of July. On that date in the forebodingly dry 1914, as I read Burroughs on Cuckoos in a sun-streaked bush recess, the ventriloquial wail of one of the little birds sounded hard by — a pretty coincidence. The shrill nature of this whistle is calculated to give its author a hearing in any bush orchestra. It is quite simple and easy of imitation. With a fluency developed in many attempts to win the confidence of other birds, I once addressed a solitary Bronze Cuckoo which, freshly arrived in the district, was placidly banqueting on caterpillars at the top of a sapling. *Cherchez la femme!* Apparently having other Spring ambitions besides the pursuit of insects, the pretty bird spread wide his tail and drooped swiftly towards the source of the invitation. But a Yellow-tufted Honeyeater had charge of the world below, and the avine Lochinvar was forced to flee from a rebuke thoroughly righteous in its vigor. A second whistle brought a repetition of the comedy. I let the matter rest at that; it would hardly have been fair to test the Cuckoo's valor (and gullibility) any further.

Incidentally, one must admit that the sex of the bird was only assumed; and, on the same principle of fairness, it should be added that there was another August day whereon a similar whistle, which much resembles the pipe of the pretty Crested Shrike-Tit, moved a *female* of that species to follow the call excitedly for a good half-mile!

In many instances the voice of a bird is a key to its nature. At all events, the call is nearly always an indication of the spirit of the bird, and in harmony with its flight. Thus, the Pallid Cuckoo flies as deliberately as it calls, and the Swift Lorikeet (to take just one instance) calls as boisterously as it flies. I see this strange honey-eating Parrot, with the aloof, gipsy-like ways, quite a lot in the early Spring, and hear it most on those grey days which may precede either rain or sunshine. Then, when "all the air a solemn stillness holds," and the earth seems to be listening tensely for the fluting of the Spring, a company of the dapper green birds with the red under-wings will suddenly start from a blossoming eucalypt, and go rushing pell-mell through or above the tree-tops, breaking the spell with their ringing, metallic "Clink-clink-clank-clink."

With customary sense of the eternal fitness of things in this way, we boys knew this bird as the "Clink," just as we distinguished its little relative with the red face by the title of "Gizzie." To the aboriginals this latter bird (*Glossopsitta pusilla*) was the "Jerryang"; but who that has heard a company of the 'Keets calling "Giss-giss!" as they hurry through the upper air, will not agree that the white boys' name is at least as fitting?

It is not to be assumed, of course, that one must of necessity go bushwards on these days to enjoy the smile of her Grace whom Hugh McCrae has prettily termed "milkmaid August." The almond trees are flowering in the towns now, and birds and bees are revelling among the snowy blossoms. The planting of trees in any situation brings its own particular aesthetic and economic reward; but there is little to excel the pleasure to be derived, at this period, from the proximity of a big almond tree to the home. Towards the end of July, when "blossom by blossom the Spring begins," the keen eye of a roving Lorikeet notes the invitation of the opening buds; and very soon then — for news of this nature travels quickly in Birdland — every branch has its bird. An almond tree in full bloom is a pageant in itself, excelling in pure radiance the magnificently assertive jacaranda and flame-tree of Queensland. And when its bird-guests are present the very air breaks into flower. There are the black-white-and-yellow of the madcap New Holland Honeyeater, and the delicate greens and whites of the "Chickowee" and Silvereye, with the distinctive colors of the Scarlet-faced and Purple-crowned Lorikeets making for a vivid impression of the Tennysonian fancy of blossom in purple and red.

I see only these two of the Lorikeets in the almond trees — the small, swift-flying "Gizzie," and, more plentifully, the aptly named Purple-crowned Lorikeet, a slightly larger bird than its congener, but hardly imposing enough to warrant the tremendous "state" name of *Glossopsitta porphyrocephala*. Happily, however, the busy little 'Keet sees not itself as scientists see it, and there is nothing but joyousness in all its movements, whether it be performing acrobatic feats in the almond tree or shouting along through the upper air.

It is worth noting, incidentally, that there is none of this clamor when the birds are in a tree close to a house. The Silvereyes may lilt away to their hearts' content, and the bees may drone unceasingly, but the little purple-crowned birds remain discreetly quiet. It may be, of course, that they are too busy with the blossoms, but, having listened to their gaily irresponsible chatter when out of sight of human habitation, or in trees that do not show up green-coated birds, I prefer to give the Lories credit for possessing an attribute which many members of a certain higher(?) genus conspicuously lack — that is, perception of the proper time to be silent!

Some years ago — to be precise, it was on the first day of August, 1913 — a small company of Lorikeets was feeding in a favorite almond tree, when a lone Butcher-Bird on evil bent dropped softly down alongside one of them. If the little fellow was expected to flutter away in alarm, anticipation was not justified. He simply uttered a reproachful cry and hopped to one side. What bluff was this! The grey marauder looked amazed, and made as though to follow, whereupon the small bird turned and screeched decisively at him. Then the Butcher-Bird hopped to a higher branch, apparently with the intention of dropping on to the plucky 'Keet. But the latter's surprising *sang froid* was too disconcerting. Presently the whole band of Lorikeets flew off unharmed, and a crestfallen Butcher-Bird sulked in a mulberry-tree.

I wonder, by the way, if this altogether attractive little Lorikeet is only of recent birth; that is to say, was it created after the advent of Bass Strait? For the fact is that the Purple-crown is the only one of the five honey-eating Parrots known to Victoria which does not penetrate to Tasmania. But then, why should the Northern States also be excluded from its itinerary? — the additionally interesting fact being that, despite its fraternity with relatives, it is the only Lorikeet unknown outside the southern portions of the mainland.

It is one thing to compare the blossom of the almond tree with other cultivated blooms, and quite a different matter to compare it with Australia's lilies of the field, the flowers of the bush. In the heart of every lover of nature these have a place apart. It may be that they do not coincide with what W. H. Hudson has called the "largeness" of the Spring mood as definitely as do, say, the gipsy voices of the Cuckoos; but, being of the earth without "earthiness," as we usually accept the term, they express the spirit of the waking bushland as not even the birds can. Without straining at a fancy, one may say the bush-flowers are the smile of the earth, which smile persuades the air to laughter, expressed in the songs of mating birds. Nina Murdoch, the New South Wales girl-singer, knows them both. In verse as delicate as Brereton's (earlier quoted) is dancing, she offers us a Perdita-gift of "orchids, green, and mauve, and white," and then tells of the bringer of this dainty posy:

> Oh! it is August, singing by the creek,
> And flitting to and fro upon the heath,
>   With busy fingers and bewitching ways
> Of darting here and there at hide and seek
> To please her babe, the Spring, who underneath
>   A leafy shelter with a wild flower plays.

It is to these little terrestrial orchids, chiefly members of the genus *Glossodia*, that the thoughts of lovers of Southern bush-flowers must go back in after years — go back with all the affection of a Briton for the primrose by the river's brink. Their rank, wildwood fragrance is potent to revive old memories, no less than their peculiarly human-like, sympathetic little "eyes." I think particularly in this connection of the blue (or mauve) *Glossodia major*, probably the most plentiful of the small ground orchids of Southern Australia.

> You are my own, of my own folk, you little blue flower
> of the Spring.

Is there, one wonders, something especially human which endears us to wild flowers of a blue color? Mr. Hudson (most widely read of British contemporary nature-writers) replies in the affirmative.

"The blue flower," he says, "is associated, consciously or not, with the human blue eye; and, as the floral blue is in all or nearly all instances pure and beautiful, it is like the most beautiful human eye. This association, and not the color itself, strikes me as the true cause of the superior attraction which the blue flower has for most of us."

If this theory were applied to the orchid alone it would be rendered additionally strong by consideration of the eye-like construction of the flower. Looking at blue flowers by and large, however, it seems to me that their attraction for us has little of the directly human appeal of blue eyes, but rests in the peaceful purity of the color and its affinity with fair weather. Accepting Mr. Hudson's suggestion, we would have to find commensurate beauty in flowers, or even leaves, of a warm brown — for who can deny the

THE GIFTS OF AUGUST

charm of brown eyes? — but we do not do so. The little blue and white orchid, then, as a widespread harbinger of blue and white days, may be ranked among the typical wild flowers of the Southern Spring.

In a somewhat lesser degree, these remarks apply also to the graceful sarsaparilla, so-called (*Hardenbergia*), which now festoons young saplings and old logs upon the hills. And everywhere about these rugged heights, hiding the scars left by roving prospectors, the pink and white daintiness of the wax-flower (*Eriostemon*) is showing strongly, while the fairy faces of a dozen less communal plants are expanding daily "petal by petal to a laugh." In that stern Spring of 1914 the particular wild-flower gardens upon which these observations centre were sadly depleted; so much so that a newspaper paragraph in the following August deplored the effects of the ravages of drought, flower-pickers, and the settlers' cattle. But the hardy little plants made a fine recovery within a few weeks, and a relieved writer very willingly tendered his apologies to a deputation of sturdy stock-owners, who took the suggestion to mean that they did not feed their animals sufficiently!

This, too, is wattle-time. August, in fact (in these Southern areas, at all events), has stronger claim than October on Kendall's tribute to a "maiden with bright yellow tresses." Particularly is this so in the bush favored by the fragrant glory known as *Acacia pycnantha*, the golden wattle. The aroma of this acacia is stronger, more expansive, than any other species known to me, unless it be that of a wattle with a whitish-yellow blossom which I have seen, and *felt*, flowering in North Queensland in the dying days of Summer.

Perhaps it is because of this heavy fragrance that

birds seldom build their homes in the flowering bushes. Reflecting on this point when strolling about a wattle-wreathed promontory of a country lake, I was delighted to chance on a nest of a Fuscous Honeyeater daintily suspended in an acacia in full bloom. It was a pretty picture; fancy might have played with the little bird as the spirit of the wattle-tree.

You may sometimes see Rosella Parrots nipping off the blossoms of the wattle, but the birds do not, of course, obtain nectar from the "dear little heads of gold." To supply this among smaller plants is chiefly the function of the heath-shrubs (*Epacris*), which now are shaking their dainty pink and white bells on the mountains. And in this case the Spine-billed Honeyeaters have a monopoly. Why? Simply because they are the only birds possessing bills long enough, and slender enough, to penetrate the delicate flowers. Personally, I have seen few more charming sights in early August than a company of the graceful Spinebills flashing in, out, and round about the colorful clusters of heath in the grey Pyrenees Range, what time their wings made a sharp, clipping sound, or reverberated rapidly as the Honeyeaters poised, Humming-Bird-like, before a flower. The picture was a beautiful living miniature, fit to rank with the magnificence of the effects created as the heavy white fog of morning melted in the noonday sun, and rose pall on pall above the valleys.

It is as yet rather early for the Spinebill to do more than "think" of nesting. About the lower hills, however, and in all bushy areas which are neither too dry nor too damp, its Yellow-tufted, Fuscous, and White-plumed cousins are already engaged in constructing their fibrous little homes, while Babblers, Tit-Warblers, Wrens, and other builders of domed nests are exceptionally active

THE GIFTS OF AUGUST

from dawn to dusk. It is an accommodating circumstance for the bird-world of the neighborhood generally, and for Wood-Swallows, Thrushes, and Whistlers in particular, that the communistic "happy families" of Babblers — the "Catties" and "Arcoes" of Victorian bush boys — are such early breeders; their large, twig-built dwellings serve admirably later on as the basis of many other bird-homes.

But we may well be doubtful whether even the Babbler is so typically August's bird as the Yellow Shrike-Robin. This bright-eyed little study in yellow and grey was very sedate earlier in the year, but now the bush is a-thrill with his agitated whisperings. In other days, too, Robin was very willing to fraternise with any human friend who called in to his bush recesses. Not so just at present. These are his busy days. Either there is a partly built little bark home awaiting completion, or a brooding mother-bird to be fed, and he scarcely has time to survey the bush-world from the side of a sapling in that funny little way of his own. Spring speeds on apace, and he is a wise individual, bird or man, who makes the most of these vital days, when the Spirit of Youth is abroad in the land.

Chapter II

SEPTEMBER REVELRY

With the passing of August there develops a steady increase in vitality among the courtiers of the Spring. September, accordingly, is nothing if not frankly exultant.

> Such joyous lips she has, such star-blue eyes,
> Such slim child-breasts and pulsing, slender throat!

Less uncertain in her moods, her smile is rather more assured than that of wilful August. She gives her secrets freely to those who will but look and listen, and you may tell the whole world if it pleases you. What else, indeed, might be expected? For the pulse of Spring beats strongly now, the alarums and excursions of Winter are less menacing, and the courtier birds and flowers are revelling in the witchery of which they are part; it is not a time for reservations.

There is, of course, no clearly noticeable physical distinction between the dying days of August and the early days of the succeeding month. And yet — well, it is easy to fill one's head with fancies in the sweet o' the year. The fact is that during many years of fraternising with the birds of central Victoria, I found myself at large in the bush on the first day of each September, and on each occasion, dull or fine, seemed to detect a new brightness in the sky, a richer tone of rapture in the chatter or songs of the birds. In sober truth, the pilgrimage became almost involuntary, a kind of festal day upon which to meander along somewhat after the fashion of Lindsay Gordon (the real Gordon, of "Bush Ballads and Galloping Rhymes"), blending scraps of his and other lyrics with the *living* music of the bush. "Some songs in all hearts hath existence" when September comes to Australia .... "Such songs have been mine."

Then there were the days immediately following. Later on, nests would be found with a prodigality almost disconcerting to one content to neglect, not to say lose, the flying hours in developing bird-acquaintances on a more parochially intimate scale. But in the early days of September the youthful bird-lover had time to stand and stare. Thus came this entry in the bush diary relating to a third morning of September:

> A peculiar day. The first clear brightness of the sun was overcome by a web of fleecy clouds, which made the background to appear as though a beautiful gauzy veil had been drawn gently over it. The tramp poet would have said again, "My fancy loves to play with clouds," had he seen these delicate streamers turning the bright blue to a soft bluish-white, some of them crossing and intercrossing each other at right angles. While gazing idly about at the foot of an old paddock I heard a Cuckoo chorus, the plaintive scale-notes of the

SEPTEMBER REVELRY

Pallid species blending with the lonely monotone of the little Bronze Cuckoo. Then, adjacent the dam in the old gully, came much entertainment from a concert of Whitenaped Honeyeaters. Dainty, jolly little acrobats! To one of them I drew very close and listened to a throaty "Joe-Joe-Joe," that was suddenly broken now and again by an irresponsible "Churr-churr!"

Follows then a record of nests found — some in new sites, others in situations known to have done good service during preceding Septembers — Shrike-Robins, Honey-Birds, and Thrushes predominating. None of these birds' nests was particularly hard to find in the district of which I write, and too often a subsequent entry in the note-book recorded a home despoiled — "rooked," in the expressive term of the Australian kiddie, "harried" in that of the British boy. Such, alack, was the frequent effect of the cause that in this month the small boy's fancy lightly turns to thoughts of nests.

For your genuine bird-lover, however, these bright days carry a deeper pleasure than even the sport of "nesting." The time of the return of the migrants and nomads has come, and he finds himself straining the hearing — and this, too, is almost involuntary — for notes loved long since and lost awhile. It does not seem to me that the voice of a bird is *necessarily* an index to its owner's disposition, but I do think that, in most cases, the rare charm of a migrant rests in its call, and all that the call suggests, rather than in the bodily presence of the bird. With one who has become accustomed to listen for these greetings year after year, the ear becomes preternaturally quick. The old familiar song or call can be detected in the distance, no matter how loud the Choirs of Spring.

There comes to mind the morning of a day in Spring when, in a pleasant recess of the bush, I tried to separate for a pair of bright young schoolteachers specific calls from out a maze of bird-melody. We soon gave it up. The rich canticle of the Rufous-breasted Whistler and the resounding trill of the Fantailed Cuckoo were acknowledged readily enough, but the finer undertones — the airy, delicate note of the Spotted-sided Finch, for instance — were entirely lost on the uninitiated ears.

Coincident with the arrival of the Cuckoos, or very soon afterwards, come the Cuckoo-Shrikes, handsome, grey-clad creatures, which have the flight of the parasitic birds. But there resemblance ends. In the view of other birds, in fact, it does not begin. They know quite well that the Cuckoo-Shrikes have none of the egg-foisting effrontery of the Cuckoos, and so the long-bodied, grey visitors are received on terms of amity.

To the roving boy of September days the best-known Cuckoo-Shrike — the Black-faced Graucalus of the ornithologist — is the Blue Jay, or Lapwing, titles which would be infinitely more acceptable than the ponderous "official" name were they not borrowed; older countries have got in ahead of us. Precedents trouble young Australia very little, however, and I surmise that *Graucalus melanops* will continue to be the "Lapwing" every time a boy sees a bird, or a company of birds, at the pretty practice of wing-waving. This, as a usual rule, is only followed immediately the "jays" have lit upon a bough. The wings then are folded, lifted, refolded, and lifted again, as though their owners are unable to adjust them quite satisfactorily. But there are times when a company of Cuckoo-Shrikes will keep this up to the extent of turning the practice into a distinct game — something in the nature of see-saw.

SEPTEMBER REVELRY

The rollicking chortle which accompanies this performance gets beyond the curiously rolling note characteristic of the Cuckoo-Shrike. I heard it to splendid advantage on one of these vital days of early September, a few years ago. First of all, there was but one bird calling, and it took a little locating, for the peculiar, indefinite trill has ventriloquial qualities. After a while, the trill was rounded off with a sharp, melodious pipe, which would have seemed to come from a distance but that it followed so smartly. Next, the musician varied matters with a human-like ejaculation, thrice repeated — something between a soft laugh and a groan. The next note, or series of notes, resembled nothing so much as the creaking of branches, and in this bar another bird joined. Suddenly, a third Graucalus materialised near by, and the three merry hearts gave a spirited trio, apparently on the little bamboo instruments from which we boys used to extract "music." I concede, however, that the birds were a good deal more melodious than, and obviously all as happy as, any holiday-making boy.

The characteristic rolling note will first be heard as the birds fly overhead on their southward journey. It is best syllabilised by the aboriginal expression "Kai-a-lora" — say that over with full play of the tongue — a name taken from the natives by Thomas Watling, who arrived in "New Holland" in 1792.*

---

* Watling, one of the earliest painters of Australian material, holds a place of considerable negative importance in the ornithology of the country, since it has been discovered that it was (probably) from his paintings that Dr. John Latham, the celebrated British ornithologist, obtained the material for describing many "New Holland" birds then new to science. In a publication of the British Museum, dealing comprehensively with the history of the collections, it is pointed out by Dr. R. Bowdler Sharpe that "in 1902 the Museum acquired from Mr. James Lee, a grandson of the famous horticulturist, of

Only seldom have I heard the call of the grey Graucalus echoing in Victoria as early as August. It is September's bird just as definitely as are the full-voiced Rufous Whistler, Reed-Warbler, and Song-Larks. Each of these birds was known to present itself during the eighth month — rarely one might stay in the South the Winter through — but then, in the colloquial phrase, it seemed to be speaking out of its turn. The guests had come, as it were, before the hosts were ready. Absurdly enough, the calendar suggested a line of demarcation. Let August "droop her deep-blue eyes in pleasant sleep," and immediately all our little world seemed waiting for the voices of these songsters.

It is, I suppose, the influence of association and reminiscence that makes me appreciate the same bird-songs far less when I listen to them in the north of Australia at other times of the year. To me they are birds of September, the elves of Spring, and I have no desire to

---

Hammersmith, a large volume of paintings executed for the latter by one of his collectors, Thomas Watling, between 1788 and 1792. The drawings had evidently been shown to Latham, who named most of the birds, and seems to have referred to the picture as 'Mr. Lambert's drawings.' They do not seem to have been Lambert's property at any time. The types of Latham's species are, in fact, founded on these drawings of Watling's. The collector was sent to New South Wales by Mr. Lee, and some of the illustrations in White's 'Journal of a Voyage to New South Wales, 1790,' were drawn by Watling, who refers to White in his volume of paintings. . . ." Then follows a list of the drawings, "as determined by Latham himself, and bearing his handwriting," together with short notes and aboriginal names by Watling.

This little Museum romance is not weakened seriously by the fact that some of its statements and dates do not square with the results of more recent Australian research. In the course of some "Notes on Australian Artists," read before the Royal Australian Historical Society (Journ. and Proc., Vol. 5, Part 5, 1919), Mr. William Dixon lays it down that Thomas Watling, according to his own story, was convicted by a Scotch jury and sentenced to transportation. He was sent out in the *Pitt*, which sailed from England

SEPTEMBER REVELRY

hear them in a strange setting. The strong carol of the Rufous Song-Lark I have heard ringing out over the wide plains of north-western Queensland in early Autumn, but the song then had not the charm of kindred bird-voices which, like the call of Wordsworth's Cuckoo, came from bush and tree and sky in the days of Spring.

Excluding the capricious little Pipit, or Ground-Lark, southern Australia knows four species of these favorite birds, namely, the old-world Sky-Lark, its Australian relative the Bush-Lark, and two distinctively Australian members of another family, known as the Black-breasted and Rufous Song-Larks. The famous introduced singer has long stayed closely about the environs of Port Phillip Bay, and is only now gradually extending along the open spaces up-country. Evidently the Sky-Lark finds the Austral bush as "weird" as did Marcus Clarke, and will not attempt to cross it. The native Larks, of course, are not so fastidious, but they, too, for the most part, prefer to live out

---

early in July, 1791. He escaped at the Cape, was re-captured after about a month, and was kept in prison for seven months waiting for a ship. Finally, he was shipped for "New Holland" on the *Royal Admiral*, which left the Cape on August 30, 1792, and arrived in Sydney on October 7 of the same year. Watling records that his employment was "painting for J. W, Esq., the nondescript productions of the country . . .," adding near the end of the letter: ". . . . My present position is chiefly owing to the low revenge of a certain military character, now high in office. . . ." The "J. W., Esq.," referred to is doubtless John White, Surgeon-General to the Port Jackson settlement. White's *Journal* was printed in 1790, and all the plates are dated December 29, 1789. Accordingly, if Mr. Dixon's dates are correct, the Museum was wrong in stating that Watling executed some of the drawings in White's *Journal* just as it appears to have been wrong in stating that he was sent out by James Lee. It seems certain, however, that Watling executed many delineations (both paintings and rough sketches) of Australian birds and flowers; and, taking the English and Australian evidence in conjunction, we arrive at the conclusion that many of our best-known birds were first named from the paintings of a talented convict.

their melodious lives in the wide spaces of the land. The Rufous Song-Lark is the exception. Given a small grassy paddock in which to feed and hide its eggs — the beauty of which, by the way, altogether flouts the protective coloration theory — its only other requirements are a few trees, fences, or telegraph wires for use as rostrums and resting-points. Frequently the bird will rise singing from the grass, but more often he — I take it that only the male bird sings — is first seen flying strongly from tree to tree across a paddock, carolling joyously the while, and continuing the song after alighting.

It may be unduly fanciful to suggest that a touch of red or rufous in a small bird's plumage is token of an ecstatic disposition; but, after reflecting on the sparkling notes and actions of the Cardinal Honeyeater and Scarlet-breasted Mistletoe-Bird, we come to the fact that the only song of Spring richer than that of the Rufous Song-Lark is that of the Rufous Whistler. "When the earth is mad with song some blue September morning," no Sky-Lark could jubilate with greater vigor than do these two wandering minstrels, newly arrived from the north, with, perhaps, the Reed-Warbler lending merry support.

The revelry of the Reed-Warbler, however, is rather of the night than the day. Certainly this modest-looking little bird sings finely on these bright mornings of Spring, but he seems to find his fullest voice on moonlit nights, or at times when the sun is overcast by rain-clouds. Perhaps it is better so. Most folk can be cheerful when the sun of Spring is shining, but there is something distinctive about a song-bird for whom clouds are an inspiration, and who, like the little god Love, accepts the rain-tears as "bright dew-drops for his wings." Here is a friend's sympathetic record of lake-side eloquence on a day in Spring:

Che, che, che, peet, peet, Georgie, tu-whee, tu-whee, tu-whee, yui, yui, yui, sweet, sweet, tinkle, tinkle, tinkle, wheree, wheree, wheree, cling, cling, cling, wit ter, wit see, now, now, now, witchee, witchee!

The translation seems to me to be no less faithful, than, and almost as pretty as, that drawn by Macgillivray from the voice of the English Thrush. I would add only that the Reed-Warbler usually lays more stress on the word "Georgie" than is indicated; so much so that when the rich voice was heard again on the early days of each September, it was greeted with, "Hello! *Georgie* is back!"

It will be noted, then, that the best of the vocalists in the choirs of Spring are, almost without exception, the travelled birds. On the other hand, the leading songsters among the birds we have always with us find their chief musical powers in the Autumn and Winter. Could any arrangement be happier than this? It pleases us, and, what is more to the point, it suits the birds themselves. The stay-at-home species are too busy now for song and play. Moreover, they are fully seized of the fact that discretion is a needful attribute of the nesting bird. Undue merriment is apt to call attention to little homes hidden away in the bushes, and these, as every bird knows, are secrets not easily kept under any circumstances.

It is for the nests of these stationary birds that the roving naturalist searches in the early days of September. There may not be news for him in every bush just yet, but it is growing day by day. If there was nothing of note in that patch of eucalypts on one day, there will be something next day — perhaps a few woven fibres where a Honey-Bird has made a start with its pendulous home, or a telltale piece of bark that calls attention to the choice of a Thrush,

Regent Honeyeater, or Yellow Robin.

There is rare pleasure in sauntering along from tree to tree on these bright mornings, peering into the recesses of bushy stumps for the round bark homes of Thrushes. The Yellow Robin, keeping to more exposed situations, can see you coming. This advantage is denied the Thrush, with the result that, as often as not, when you part the bushes growing about a scarred old stump, you come face to face with a shapely grey bird, looking at you with big, bright eyes, half-frightened, half-fearless.

A mother-bird is, I think, rather less afraid under such circumstances than when she returns from foraging and finds a stranger looking curiously in at her treasures. Over many years I can still hear the shrill, musical shout of alarm emitted by a mother Thrush who caught me peering at her trio of helpless babies. The nestlings were quite blind, but at the sound of the mother's voice every neck was extended and each wide yellow mouth opened in a plea for food. Evidently the hearing of a baby bird comes at birth, whereas ability to distinguish the *nature* of a call does not develop until the eyes open.

The true melody of the Thrush is rather of the Autumn than the Spring. Now the notes are flung out, not with the reflective ease of the retrospective vein, but with greater passion. Once I heard an excited Harmonica whistling spiritedly while on the wing, after the manner of a Butcher-Bird, and on another occasion one of the handsome grey birds caused surprise by chuckling indiscreetly while nest-building fifteen feet up in the fork of a sapling — a young wife, I suspect, unable to repress her exultation in the marvellous tale and the beauty of the Spring.

The nest of the Thrush, in the bark material and general structure no less than in general situation, much resembles that of the Crested Bell-Bird. No birds'-nesting boy could ever distinguish between these homes until the eggs were laid, after which it was easy to separate the black-spotted white eggs of the *Oreoica* from the rich-brown-spotted treasures of the Thrush. The prettiest eggs of the kind I ever saw were a complement of three housed in a novel situation, the bank of a bush creek, to wit. The Thrushes have no burrowing powers, but they had nicely cleaned out an old hollow in the steep, dry bank of the streamlet, and woven bark therein — a neat nest, and quite out of sight. But the keen eyes of boys found it, and the three eggs were taken. Then the Thrushes built again, at a point only six yards from the old site; this time the nest, though very cosily made, was visible from the opposite bank. It was found by the boys, who showed it to me; and certainly they deserved commendation for forbearance, the eggs being the most beautifully marked trio I had seen among very many Thrushes' homes.

In another of these capricious birds' nests under observation at that time, from which the three young ones got safely away, the usual material of bark was discarded, and the structure built of grasses, woven into the cleft of a stump, and plastered together with mud.

If it is not the nest of a Thrush that rewards your fossicking about the bushy stumps in September, the chances are that a brooding Regent Honeyeater will be disturbed. This pretty sprite in yellow and black has given over now its Winter revels among the blossoms. There are more serious things to think of, and the beautifully melodious voice is not often heard unless a mother-bird be flushed from one of the neat bark homes — a cross between the nests of the Thrush and Yellow Robin — tucked away amid the screening bushes of a stump.

Strolling through the bush on a bright morning in early Spring, my roving eye was caught by a few tell-tale shreds of bark protruding from a bush-crowned stump only a few feet in height. As the footsteps approached, off flew a Regent Honeyeater. There was one young bird in the nest, and the cheeping of this babe roused the parents to a high pitch of excitement. First they came themselves, darting at my head, and chattering in musical anger. Then they brought numerous relatives, and the whole company set up a hue and cry that continued until the intruder departed, upon which they *cooed* softly together. It was an altogether mellifluous chorus, expressively embodying a mixture of fraternal thanks and congratulations. "We hope," those sweet-voiced parents plainly said, "to do as much for you all some day!"

I am reminded here that in the case of other of the nesting Honey-Birds of September the voiceful protest at human intrusion is reinforced by a clever artistry. The Yellow-tufted and White-plumed species have not the melodious voice of their Regent relative, but each of the little creatures is adept at feigning to be wounded when its treasures are menaced. John Burroughs restricts this ruse in America to birds which nest on the ground. There is no such limitation in Australia. Several of our ground-dwelling birds (notably the Babbling-Thrush) will flutter off their nests and drag themselves along the earth when danger threatens, but none is a more consummate actor than the Yellow-tufted Honeyeater. Slipping out of its dainty cradle suspended in a sapling, this pretty tragedian will go tumbling and fluttering along over the roughest of ground, beating its wings distressedly, and screeching meanwhile as though in mortal agony. So soon, however, as the uninitiated or indulgent visitor is drawn from the danger zone, the tragedy gives place to comedy — the actor shakes its little tail in the shelter of a friendly tree!

Why, one wonders, is the exercise of this pretty device restricted to non-migratory birds? Is it because of their strong voices that the wandering species rely on vehement protests, rather than subtle devices, when their nests are apparently in danger? There is one other small bark home I can always be sure of finding on September days. Like the Thrush and the Regent Honeyeater, the Yellow Robin loves to keep as close to mother earth as safety will permit. Having no fondness for the screening of bushes, however, it must needs build its house with more neatness, not forgetting to have an eye to harmony of color in the site. Moreover, as *Eopsaltria* has not the strength of voice of its larger neighbors of the bark homes, it has recourse to stealth. Sitting watchfully on the nest, Robin will always see before being seen. As the danger grows, she dives quickly and flies softly, flush with the ground. The progress of events is watched then from the side of a handy tree, and, should there be callow young in the nest, possibly the little mother will reappear on the scene as a fine imitation of a disabled bird. She does not tumble and scamper along in the style of a distraught Honeyeater or Bush-Chat, but "fluffs" her feathers and hops about, the effect being suggestive of a sick, rather than wounded, bird.

But that initial dive is undoubtedly the move upon which most reliance is placed. A philosophic Robin was one day returning my interest in her lack of movement when she caught a warning shout from a sentinel Honeyeater. The little head turned quickly to one side for a glance skyward, and, almost in the same movement, she dived quickly low into the undergrowth. Shortly afterwards a Goshawk came beating steadily over the tree-tops. What pretty fraternity of interest prevails in Bird-land in time of danger!

For more than one reason, both economic and æsthetic, it is well that the stay-at-home birds, knowing every niche of the woods, have got well on the way with their housekeeping ere the travelled species begin to spy out the land. And not the least important of these reasons is that the roving naturalist is enabled thereby to keep pace with most of his appointments. Think, ye bird-lovers, how bewildering it would be if all the bush-birds nested at one time; and then give seemly thanks for a happier arrangement!

Chapter III

OCTOBER THE WITCHING

Once upon a time, so the old school-book had it, there lived an ass which found cause for discontent in all seasons of the year, and ever was longing for change. In the cool of Winter he yearned for the fire of Spring; increased work in the warmer months turned his thoughts to the Autumn; and so on, until, finally, this strangely human-like philosopher awoke to the conviction that it was well to make the most of whatever joys were to be found amid the pulsing process of the seasons.

The excellent moral of that little doggerel-verse story was not lost. In one case, indeed, it deepened into a pleasurable puzzle as to which quarter of the year was really the most gracious; and then — getting down to "sub-species" — as to which month of the Spring had the greatest claim on the affections. Paris adjudging the Graces!

Hosea Biglow, kin-spirit of the lilting Bobolink, made the absorbing discovery that the American Spring "gits everything in tune, an' gives one leap from April into June." It is not so in this land of the sun. The spirit of the Australian Spring is not less thrilling, but is more deliberate — shall we say more benignant? Such sudden, assertive prodigality of joyous gifts as those conveyed by genial "Hosee" is liable to become embarrassing. *Our* Lady of the Spring has sudden little ways of her own; but, in a general aspect, she makes up her mind with commendable tact.

That decision may be made in July. In the following month there are agitated whisperings among the birds and bush flowers, and so the renascence goes gently on and on — through wilful August and splendid September — until, in the witching month of October, the laughter of bird and bloom is at its merry height. The promise of Spring is fulfilled.

But, mark you, if that withdrawal of the veil of Winter is gentle, it is none the less sure and steady. Twelve years or so ago a Victorian bush-boy spent something over a month in Melbourne. September came to the city. There was new life in the air, a subtle but perfectly distinct breath from "where the good winds blow . . . over the hills and far away." Soon began an interesting little conflict between city and bush. Had there been any doubt as to the issue, it would have been decided by the recollection that, "up there," the Rufous Whistler, newly arrived from the North, was pouring out its exultant heart to speeding September. What siren voice of the city has the potency of that call?

> There were myriad lights on the great white road,
>   And a voice that called me, "Stay!"
> But the inland breeze whispered down to me
> With the sound of trees and a melody
>   That the breeze and the grasses play . . .

OCTOBER THE WITCHING

It was then, when the arms of the bush opened to receive back the youthful Nature-lover, that vivid realisation came of the sure steadiness of the onward march of the Southern Spring. September having been lost, it verily seemed that the spirit of the sweet season had lifted the bush-world straight from August to October. The migratory and nomadic birds had returned to swell the new-old chorus, the gold of the wattle had been transferred to the everlastings of the hills, the "billy-buttons" of the roadside, and the carpeting Cape-weed of the paddocks; the dainty blue "pincushions" created a pretty border to the bush railroads, and every bush-orchard held a riot of lush grass and fragrant apple-blossom. It is no exaggeration to say that the luxury of it all — melody, visible beauty, and fragrance — was literally bewildering in the suddenness of its salutation; and, though one may not affirm that the loveliness increases with the passage of the years, it is certainly true that the vital experience is a poignant memory — can still distinctly be felt.

As the blossom of the almond-tree is to August, so are the fragrant flowers of pear and apple to October; indeed, they are as much a part of this lavish month as are the dainty *Diuris* orchids and other modest wildflowers. Your bush-orchard, particularly if it be not too precisely trimmed, is always a pleasant spot; and just now it is a source of sheer delight, something to exult in.

I recall with what pent-up exuberance a somewhat dishevelled old orchard was greeted by a class of junior school teachers whom I led abroad on a day in October of 1914. Our Lady of the Spring seemed to have got all her arrangements awry in that fateful year, and the intervening bush was dry, parched, and strangely inhospitable. Under those circumstances, was it surprising that the young people greeted snowy blossom and green grass with all the gratefulness of travellers at an oasis?

Beautiful under the lancing rays of the morning sun, the old orchard is a hymn of delight at evening, when the cooler air draws a richer fragrance from the flowers, and the gathering dusk evokes the lirruping lullaby of Madge, the solemn merriment of the Kookaburras, the reflective "sweet pretty creature" of the Wagtail, and the calmly resigned semitone of the Ishmaelitish Pallid Cuckoo. And what of the night, when the Puck-like spirit of October, revelling in the old orchard, "sprinkles its buds, in beneficent floods, to gladden the way of the moon?" Surely a fit place (the more so if there chance to be vines and pomegranates) for our Hellenistic Hugh McCrae to foregather with the shade of his beloved Herrick, there to watch stars flutter through the blossoms:

> . . . . till the still
> Deep moon lay down her bowl of silver light
> Below these lilies near the orchard hill.

As a matter of course, October is the month when birds resort to the orchard for nesting purposes. The whistling Whitethroat of the rufous breast is there in September, but it is not until little leaves are twinkling on the erstwhile bare boughs that the Whistler, the Wagtail, and the sweet-voiced Goldfinch can find the shelter necessary for the safety of their little homes. This pretty imported Finch is apparently constant to gardens and parks for home sites (though its fellow-Britisher, the Blackbird, is now taking more or less kindly to bush trees), but Wagtail and Whistler are not at all averse to nesting in other situations of moderate convenience. I have even known the familiar little black-and-white bird to build at a height of fifty feet on a slender tree-limb in a back yard. Frequently the pretty nest is situated on a branch overhanging water; but there is no spot which harmonises so well with the soft grey of the nest (or *vice versa*) as the shady limb of an apple-tree.

"The Wagtail's nest is beautiful," ran a childish essay: "with the eggs in it is *more* beautiful; with the fond mother sitting on the nest it is *most* beautiful." Superlative exhausted, what would that child have said of the picture presented by a Wagtail at home amid the apple-tree's wealth of October blossoms?

It were easy to enumerate several other species of birds which are more or less constant to the bush-orchard for nesting quarters, but none of these is so much a part of October as the two travelling species of Wood-Swallows, and the spruce study in black-and-white known by the cumbersome title of White-shouldered Caterpillar-eater. Of the three common varieties of Wood-Swallow, the dusky species (*Artamus sordidus*) is constant to the south the whole year through. This is the philosophic brown bird which is chiefly notable for the communistic tendencies it develops in the Autumn, when as many as fifty of the "tribe" may sometimes be seen clinging together in the fork of a rough-barked tree, in a manner vividly suggestive of a cluster of bees at swarming time.

The other two species are the White-browed and Masked Wood-Swallows (*A. superciliosus* and *A. personatus*), two handsome companions which are wont to reach Victoria from their northern sojourn in October of each year. It is this practice that makes them to share with the larger Cuckoo-Shrike the colloquial name of "Summer-Bird," while the soft grey coloring of the birds' backs is responsible for the second local title of "Bluie." "Bluies" they were to us boys in the bush days of old, but even the charm of reminiscence will not allow that name to eclipse one which a school-class volunteered in later years. "Skimmers," those kiddies called the graceful birds, and the title is almost adequate — fit tribute to the essentially graceful aerial skating at which these two species, of all the Wood-Swallows, are particularly adept.

The wonder is that the beautiful soaring flight of these "Summer-Birds" beneath the blue heaven of October has not caught and held the fancy of Australia's poets in the way the Bell-Miners captivated Kendall, the Wagtail James Thomas, and the wild Black Swans Paterson.

The "Skimmers" may always be looked for during the opening days of October. If they come either before the first week or after the second, it may be assumed that the season is somewhat out of joint. Thus in 1914, which (as observed earlier) was forebodingly dry and hot, the White-brows astonished us by showing up on September 18. But, as though conscious that the blazing blue skies had led them to make a miscalculation, they disappeared again immediately, and were not further reported until October 7. On that date the bush was suddenly seething with the pretty chestnut-breasted birds of the white-brow, and with them this time were many of the black-faced, white-breasted "Masked" species. Within seven other successive years of which records were kept for the district, only once did these gauzy-winged wanderers misjudge the opening days of October; that was in 1912, when a mixed advance guard of about sixty members of the two species came down with a high north wind on September 29.

These exceptions do but prove the rule. Our bird-world knows no arbitrary law in the matter of calendar months; but, for all that the Australian months merge one into the other almost imperceptibly, the migrants and the nomads are able to "take the sun's height" as accurately as any American bird. And none displays more consistency than the "Skimmers."

For the first day or two of their arrival in Victoria the Wood-Swallows are not perceived by the unobservant. They are, indeed, almost unbodied voices, keeping as they do at a tremendous height in the air, whence their combined chirping drifts down to the watchful ear as the final promise of Spring. Presumably they come to the trees to rest on the night of their arrival; but, if so, they must go aloft again at dawn, for it is only by degrees that the land is spied out to their satisfaction.

But when the aerial evolutions have been given over for the time being, what bustling and clatter there is in the Wood-Swallow world! In one day the byways of the bush are filled with busily nesting birds that were not there on the previous day, and practically every one of the finer bird-calls is lost in the harsh, chiding chatter of the "Bluies." In the case of each species, the nest is run together very rapidly; these companions of the air have little time to spare for artistic home-planning, and their fragile platforms of fibres, small sticks, grass-bents, or small pieces of bark, soon accommodate pairs of quietly colored eggs (maybe three), which are hatched quicker than those of most other small birds, and in record time the insect-nourished young have grown to the flight stage. The Cuckoo that fancies a Wood-Swallow as guardian for its intrusive egg has to reach a quick decision on the point; the "Skimmers" give little opportunity for reflection on the advantages and disadvantages of the situation.

The Wood-Swallow is probably the most accommodating of Australian birds in the matter of nesting sites. It is content with practically any point of vantage that will give purchase to the scanty platform — hedge-row, prickly bush, cleft in stump, the deserted nest of another bird, protruding piece of bark, dry bush, fork of a gum-tree (at any height from two to thirty feet), fruit-tree, or even the slightly hollowed top of a fence-post. Very little attempt is made at concealment, but even if anything of the kind were practised, it would be rendered negative by the indiscreet rebuke administered by the birds when the nests are approached. Obviously, the only safeguard to the propagation of the species is the speed with which the broods are reared.

In October of 1913 the "Skimmers" under discussion (I am assuming that the birds return to the same district each year) made an unusual departure by showing a decided penchant for nesting in green bushes — that is, among or near the actual leaves — a most remarkable procedure for birds other than builders of such cleverly woven nests as those of the Honeyeaters. One pair of Masked Wood-Swallows went so far as to try to emulate the fine-billed Honey-Birds by making a clumsy attempt to *suspend* their nest. The reason for this curious departure, which was followed by approximately nine-tenths of the visiting "Skimmers," never became apparent; I know only that the experiment was not attended with success, most of the nests being despoiled.

For all his noisiness, the male of each species is quite a model helpmate. He takes his turn at the brooding, and, moreover, is always willing to sit by and cheer the female with a little whispering music as she attends to the home. Until a few years ago I had thought that the more placid, less assertive Dusky Wood-Swallow was the only member of the genus in the habit of dropping his voice to a confidential warble when sitting by the nest. But there came an occasion when, as I sat waiting for a Bell-Bird to return to its bark home on a stump, a male White-brow from a nearby nest gave over chiding harshly, and broke into a run of soft, prattling notes. In these could distinctly be detected the "pee-pee-pee" of the Brown Tree-Creeper and the long-drawn "wee-wee-woo" of the Pallid Cuckoo. Was it mimicry? I hesitate to say. We are too apt to generalise in these matters, and probably many birds are given credit for powers of mimicry when the notes are purely natural, only allowed out of the music-box on rare occasions. Otherwise, Australia must surely be replete with bird-mimics.

The White-shouldered Lalage (save us from such a mouth-filling name as Caterpillar-eater!) has a good deal in common with the summer-loving Wood-Swallows. It delights to follow the chariot of Phoebus from one end of Australia to the other, and, while it is considerably less reliable than the "Skimmers" in regard to time of arrival in the South, it may fairly be ranked with them as a bird of October. How often of old, in the brightness of the "bird-month," have I listened with delight to the sparkling, Chaffinch-like chatter of a newly arrived Lalage (*i.e.*, prattling-voiced), and seen the handsome male bird singing on the wing!

This shapely study in black-and-white, along with his quietly garbed wife, usually arrives in Victoria towards the end of September or early in October. Sometimes they may not appear until November; and then, again, they may not appear at all. I have found the same "glorious uncertainty" to prevail with the species in Queensland. There the larger Pied Lalage, a bird of the brushes, which has never been reported as a Victorian visitor, may often be seen in the cooler months; but its little white-shouldered relative is missing then, and only makes spasmodic appearances in the Spring. Thus, in the Spring of 1916, the bird of the chattering song was very plentiful about Brisbane; and in the following Spring, with no apparent alteration in conditions, there was not one to be seen.

In my experience, the best year for Lalages in Victoria (of recent times) was 1910. The sensitive birds were in no way misled by the passing menace of heavy snowstorms which occurred in various parts of the State as late as the second week in October, and, on the beautiful mornings towards the end of that month, I found them nesting freely in the *Pinus insignis* of a public park. The only safe period to fraternise with the birds in that area was shortly after sunrise; for the Nature-study movement has not stamped out the primal collecting instincts of *all* small boys, and the trouble was that the ecstatic Lalages did enough towards betraying their nests without any human assistance, however unwitting.

It is a rash bird — one whose joyousness overcomes its discretion — that sings while on the nest. Were it not for such amiable lapses, and the fact that the bright-colored male bird takes part in the brooding as well as the building of the home, the nest of this species would rarely be discovered. Neatly woven of soft, fibrous material, it has much more of protective coloration (particularly when placed in a fruit-tree) than those of the Wood-Swallows, and, moreover, it is so remarkably small that the wonder is how the babies manage to find sufficient accommodation. The eggs, usually two in number, are much like the pretty, green, brown-spotted eggs of the Yellow Robin.

In that same year (1910) I had three nests of this cheery prattler under observation in an old bush-orchard. One was built in a cherry-plum tree, on the identical site where a brood had been successfully reared in the previous Spring; from which it may fairly be deduced that these roving spirits, like many nomadic and semi-stationary birds, will remain faithful to a nesting-spot where they have been treated with something of the consideration which is their due.

Jacky Winter at Home. *(Photo. by J. H. Foster.)*

Jacky Winter's Rostrum.
"The big dry tree by the roadside."

"The madcap New Holland Honeyeater." *(Photo. by L. G. Chandler.)*

August Purity. Honeyeaters' Home amid Wax-flowers.

Heath-gatherers on the Mountains.

A Guest of the Heath-bells. "The Graceful Spinebill."
(Photo. by R. T. Littlejohns.)

"Double-decked" Nest of Tit-warbler.

Yellow Robin brooding. "Typically August's Bird."

A Lesson in the Bush. Trees denuded of leaves by hail.

A September Outing. Trainee Teachers.

Nest of Babbling-Thrush.
"A ground-dwelling bird."

Nest of Rufous Song-lark.
"In a grassy paddock."

Nest of Grey Thrush.
"In the recess of a bushy stump."

"Georgie," the Reed-warbler.
(Photo. by D. W. Gaukrodger.)

White-shouldered Caterpillar-eater.
"This shapely study in black and white."

Bird and Blossom.
Wagtail on nest in flowering apple-tree.

Wood-swallows' Nest.
"Even on a fence post."

English Blackbirds' Nest in eucalypt.
"Taking kindly to bush trees."

Wood-swallow at Home.
"The promise of Spring is fulfilled." *(Photo. by R. T. Littlejohns.)*

Wood-swallows' Nest in an Orchard.

Caterpillar-eater at nest. "He takes part in the brooding."

Female Caterpillar-eater on nest.
"His quietly-garbed wife." *(Photos. by D. W. Gaukrodger.)*

Nest (opened out) of Spotted Diamond-birds. "They breed in hollows."

Tree-creepers' Home. "In an old kettle."

Well-behaved children. Baby Yellow Robins.

Baby Crested Bell-bird.
"The restless little creature."

Female Red-breasted Robin brooding.
"It's a pity that poor Jenny is so plain."

Nest of Red-capped Robin.
"A neat little structure."

Australia's Genius — Kookaburra and young.

Hungry baby Red-caps. "Will they never come?"

Chapter IV

THE PASSING

"I wish it were always Spring!" A trite sentiment, this, but notable in the present instance because it comes from a British bird-lover, a man of attainments in literature and the lore of Nature. The expression surprises by reason of its author, and I wonder idly if he really means, or meant, what he wrote. For surely a little reflection will show the sentiment to be as weak as it is superficial.

The Springtide of the year, as with the Springtide of life, must ever be the playtime of the earth, but no less fundamental a principle is that of change. I hold, indeed, that the comparatively even nature of the Australian climate is one of the chief factors in the poetic product of the land; its geniality conduces to song, but there is not the sternness and vitality that beget world-music.

Of what avail to reflect on the grotesque and staggering possibilities that suggest themselves in the idea of a mundane world minus the pulse of the seasons? Let it be said, however, that a Nature-lover comes close to breaking faith with both his old Nurse and himself when he pines for what is not in a matter of this kind. There is felicity and interest, even mirth and instruction, for him in the whole of the varied moods of the changing year; and, what is more, he breathes the deeper for a temporary freedom from the imperious mandates of the wanton lady of the blossom-face. And there is nought to fear on the score of faithfulness; the flowers will bloom in other years, and "old songs with new gladness" are assured when the wandering minstrels return in the fulness of time.

November in Australia marks the dying days of Spring. The time for ostentation is past, but there is, generally speaking, a clear divergence between the last week of October and the second week of the succeeding month. Now the metaphorical fire of the rising Spring has given place to a literal heat, which approaches at times 100° in the shade. Bursts of rapture from mating birds are rarely heard at this period, and in their stead rises the chant that has caused a cessation of the melody — the incessant supplications of baby birds.

November, indeed, in southern Australian bird-lands at all events, is pre-eminently the period of the youngster. Wisely enough, the builders of warm, domed nests almost always have their offspring on the wing before the heat-waves begin to dance across the landscape; most of the other stationary, early-nesting species are tending second or third broods, and the majority of the migrants are also busy with the cares of the household. Of course, there are still to be found bird-homes containing the promise of life — see how resentfully small birds are still chasing Cuckoos! — but for one nest with eggs now you find half a dozen either containing babies or deserted, only a crumb-like litter remaining at the bottom of the nest to tell that the brood has got safely out into the big world.

For the most part, the birds still busy breeding are those which build in well-shaded situations (particularly the imported Goldfinch, Greenfinch, Blackbird and Thrush) and species for whom the heat of the sun holds no terrors. Some little Australians enter into both of these categories. The tiny Diamond-Birds, for instance, the Tree-Creepers, the rainbow-hued Bee-eater, the Dollar-Bird, and many of the Parrots — all of these sun-lovers are not only adaptable to dry conditions in the matter of food, but they breed in hollows. And so there is no call for any of them to sit with protecting wings outspread above a nest, as many less intuitive parent-birds have to do, shading callow nestlings from the menacing sun of November. A friend writes me from the dry west of Queensland that practically the only bird to be found breeding there during the bad drought of 1919 was a species of Tree-Creeper, young birds of which species were a feature of the locality.

The Brown Tree-Creeper, the "Wood-pecker" of roving boys, and *Climacterus scandens* of the ornithologist, is an old friend. One of the outstanding recollections of boyish days in Birdland is that of peering into a stump, hollow with age, for a glimpse of three beautiful, pink-spotted eggs. (The Tree-Creepers are among the few birds to break the law under which birds which nest in hollows lay white eggs.) Essentially faithful to a favorable locality, the Brown Tree-Creeper will return year after year to any situation that has served well, be it hollow limb, fence-post, stump, or any old receptacle of moderate convenience. I have known "Wood-peckers" to cling steadfastly to a favored orchard, impartiality being shown only in the choice of a situation; now it was the cleft of a battered kerosene tin balanced precariously on a fence-post, then a sordid-looking jam-tin suspended against a shed wall, and again an old kettle.*

Possibly it was this same stay-at-home pair of birds which, in earlier times, achieved many families in the moderate peacefulness of a prosaic old fence-post lower down the gully. The first recollection of that post and its tenants dates back to "nesting" days. Continual harrying of the birds led to the post being vacant for a few years, but during that time force of habit, or anticipation, kept me peering into the darkness of that suggestive hollow. Ultimately, there came a Springtime when the comfortable old post was no longer "to let;" and for as long as I knew it thereafter it was occupied by the plump brown birds with the orange flight-marks on the wings.

It is a feature of the home-life of the Tree-Creeper that, no matter how deep the nesting-hollow, the sitting bird seems to possess an uncanny power to detect danger; and usually it is well away by the time the nest is visited, piping a shrill invitation at the entrance to an *unoccupied* hollow, nearly always, as many a deluded small boy has found, in a situation which necessitates a stiff climb. But there came a day when, after the flushed bird had returned reassured, I approached the old post quietly, and gazed directly down into a pair of clear, timid eyes. The mother came forth again with a startled rush when the menacing presence was withdrawn, and pleaded for her three callow babies from the side of an adjacent tree.

Those droll youngsters were less sophisticated. A slight scratching or whistle at the top of the post olater occasions was sufficient to persuade them to balance precariously and comically on their tails and screech

---

* Just as this book goes to press there has come advice that the Treecreepers are nesting again in the kettle for the ninth year in succession.

for food. But the indiscreet anxiety of the parent birds never relaxed. I came upon them, subsequently, giving the fledglings lessons in locomotion, and the resultant commotion disturbed the placidity of the entire gully. As the "Pee-pee-pee" of the Tree-Creepers grew in intensity, an unusual thing happened — a couple of young Fuscous Honey-eaters jumped out of their cradle hard by. I replaced the smaller of the pair, but it was thoroughly alarmed, and would not stay. Moreover, the cheeping of the baby Honey-Birds drove their parents almost frantic; they joined in the tumult, at the same time fluttering and tumbling along the ground, and generally presenting the appearance of birds in the last stage of distress.

If there is one baby bird-voice more prominent than another at this period in the district of which I write (the fringe of the Mallee country of Victoria), it is that of the Harmonious Shrike-Thrush. The adjective is used only to denote the species, for the voice of the Thrush at this stage is anything but harmonious. It is a piping, querulous monotone, more in accord with the rising stridulations of the cicada than with the full, rich whistling of its bright-eyed parents. By the same token, it is a singular thing that the sweet-voiced adult Grey Thrush, so far from being always harmonious, at times utters a succession of screaming notes suggestive of one of the Parrots. I had, in fact, ascribed the bar to the familiar Rosella Parrot until a Thrush was seen, on a day in the late Spring, uttering the wild, half-sane shout; and endorsement came from John Burroughs' record of a similar trait in the disposition of an American Thrush.

That scream, I suspect, is only emitted when the bird puts aside the spotless traditions of the gentle Thrush clan, and gives play to the Shrike side of its nature. Similarly, it is probably under an hereditary (Shrike) impulse that the soft-eyed *Harmonica* occasionally despoils the nests of smaller birds of either eggs or callow young. In one of the old orchards upon which these notes chiefly centre a sturdy bush woman was engrossed one day in bewailing the many little homes of her bird friends which a strangely degenerate Thrush had wrecked. (Babblers and Honey-Birds, she related, had been forced to combine against the marauder.) Suddenly one of the handsome grey birds flew to a tree close by, clapped its wings imperiously, and fluted with clear emphasis: "Dick, Dick, Dick's a *pret*-ty boy!" The suggested words were accepted with a quaint seriousness. "Pretty boy be blowed!" rapped out the offended woman. "I'll 'pretty boy' you if you rob any more nests!"

The rich, sweet enunciation of those Thrush words are characteristic of the bird. To particularise just shortly, there are the sweet, chuckling melody of Autumn, the four-note, intimate call of invitation to the nearing Spring, the challenging tumult of mating-time, the loud, single call of alarm when danger menaces the nest, and the piercing scream of odd moments. But no one bar or note is so generally used as that quoted in the words, "Dick, Dick, Dick's a *pret*-ty boy." It has got into Australian poetry, not in those precise words — for human interpretations will vary greatly — but in a dainty plea for dalliance:

> A wild Thrush down where the flood waves rolled
>   Pipes to the water, the blossoms, the day,
> To the passing pageant of blue and gold:
>   "Don't, don't, don't, don't, don't-go-away!"

Very few Australian bird-notes, indeed, have so lent themselves to human transcription as those of the neighborly Thrush — this soft-eyed creature whose friendly ways have caused it to become known as "the woman's bird." When the "Grey Thrush in the wattle tree" called "Oh, you pretty dear!" in the hearing of "Jim o' the Hills" (Dennis), the bird was obviously in its sweet Spring voice. And many another merry, stuttering phrase, full of good conceit, may be gathered from the sympathetic interpretation of Miss C. B. Coutts, a lady to whom I am glad to render acknowledgment as the part-director of these fancies at the old school. Her "Bob-bob-bob-bob, White-cap!" is a close transcription of another Thrush bar (akin to that in Miss Cole's verse, quoted above), and the last stanza of her poem is quite in the spirit of the rich song of the bird:

> Bravo, bravo, gay bewitcher!
> All the countryside is richer
>   For your lavish joy this morn in merry stave;
> "Wh-wh-what joy! More joy!" Hear him!
> Who could be despondent near him? —
>   Such a perky, chatty, cheery little knave!

In these warming days you hear but little of the golden voice of the Thrush's confrere, the Crested Bell-Bird. The round, full tolling no longer echoes through the aisles of the bush as it did in the cooler months; for the Bell-Bird now has its Thrush-like nest to tend in the recesses of a bushy stump. Dry weather troubles this queer bird but little; in fact, the foreboding Spring of 1914 gave me more nests of the species than I have ever found before. Not by the most careful subterfuge was it possible, however, to get on terms of amity with those strange birds. Parents refused to return to their homes in face of an intrusive camera, and callow babies were no more amenable. One trio of nestlings closed their eyes and waved their heads in precisely the fashion of outraged caterpillars, and the single inmate of another nest, although not nearly able to fly, jumped out at a touch, and hopped away as fast as its baby feet could carry it. This action quite met with the view of the parent bird, which hopped along the ground ahead, calling its offspring in a confidential chatter. Brought back to the nest, the restless little creature repeated the pretty indiscretion again and again, and had perforce to be allowed to go its own sweet way.

For a half-fledged bird to desert its cosy nest on slight provocation is no more usual than it is natural. The case of well-feathered babes is different. Tiny wings may have been beating the air for several days without their owners stirring one-feather's-breadth from home; then, impelled by sudden fright, they will fly strongly. The most notable instance of the kind recalled at the moment concerns a family of the pretty black-and-white Lalages. In the heat of a November day three young birds sat straight up in their tiny nest — possibly so for comfort, but probably because in that position their little grey bodies resembled closely the branchlets of the apple-tree in which the nest was built. And, as soon as a fraternal hand approached them, the three youngsters took their first flight. In most cases this is fluttering and shortlived — "creep afore ye gang," as the Scotch lullaby has it — but these three flew strongly, one (obviously the first-born) sailing right away over the tree-tops.

With the Bell-bird and Lalage (Caterpillar-eater) bracket the White-browed and Masked Wood-Swallows and Red-capped and Black-and-White Robins, and you have a half-dozen pretty birds which may be found nesting in the forest areas of central Victoria in the hottest of Novembers. It all gets down, I suppose, to a strictly materialistic consideration, *i.e.*, that the particular insects favored by these birds are more plentiful in a time of drought. The sorry thing is that none of them turns attention to the larvae of the slug-moth (*Doratifera*), a brightly colored, stinging caterpillar, whose fondness for gum-leaves causes serious losses to eucalyptus distillers in dry seasons. On various occasions in the late Springtime I have seen young forests looking almost bare and stark through the ravages of these caterpillars, the "stingarees" of the bush-boy, whose only claim to favor is that they have a wholesome influence in restraining the tree-climbing propensities of the said boy. I have seen a Bell-Bird's nest in a tree swarming with these repulsive caterpillars, but none of them was included in the decorative scheme of the nest, which contained only grass caterpillars.

As for the Caterpillar-eater itself, the bird is no more entitled to the name than a host of other species, and it is reasonable to surmise that it got this vernacular title because nothing else was offering.

There is no cause for wonder in that the Wood-Swallows and the Lalage should come out of the interior at certain times; they are birds which follow Summer merrily. But it *is* a curious thing that two particular Robins, birds which wintered in the coolness of the fields, should choose to breed on hot, scrub-clad hills or plains, what time their close relatives — Flame-, Scarlet-, Pink- and Rose-breasted Robins — are spending the summer in the benevolent shade of mountain gullies.

The flitting Red-capped Robin is the gem of my dry hills in the dog days of Spring. You hear an airy, quavering warble, suggestive of a child's gelatine rattle "played by the picture of Nobody," and betimes, if you are lucky, there will appear a dandified little Ariel, who carries his garb of red, white and black with all the easy grace of the genuine aristocrat. Changing the social status of our subject, it is the scarlet Liberty cap that separates the species from the Scarlet-breasted (white-capped) Robin, the regal little bird in whose fragile notes our humorous poet caught the words: —

Dear, it's a pity that poor Jenny is so plain!

As with the Scarlet-breasted Robin, the little "Jenny" of the Red-cap is quite plain. Sometimes, indeed, the paternal bird himself will be met with in the same brownish garb; that is to say, he will breed while yet in immature plumage, which possibly takes two years to merge into brilliance.

Towards the end of the well-remembered Spring of 1914 I was poking about a hillside which appeared to be scarcely hospitable enough to entertain any bird-guest, when the appearance of an important-looking female Robin, and the exercise of patience, led me to the discovery of a nest — a neat, protectively colored little structure, placed in a fork of a slender eucalypt. No male bird was to be seen, but an experimental twitter brought the fightable little fellow dashing up; his plumage was brownish, with just a "faint fresh flame" showing on the head and breast. Again the familiar rattle drifted airily in, and presently appeared two other male Robins, with breasts of that precise shade of vivid red borne by the scarlet bottle-brush. From the fact that these two beauty-birds were on terms of amity it may be deduced that each was soberly mated; a month or so earlier the kingly wee creatures, impelled by hearts as fiery as their breasts, would have been disputing fiercely, if harmlessly, over one demure "Jenny" Robin.

A good observer states that in the dry north-east of Victoria little Red-cap, who is often known as the Mallee Robin, breeds in August and December; but the only months in which I found nests further south were October and November. It is a curious fact that few other nests are so cosily lined as this — curious because it would seem that a plentitude of feathers and other soft material would be more suitable to an early-nester such as the Yellow-breasted Robin. Reflecting on this point — the negative value of blankets in warm climates — I have sometimes wondered whether little Red-cap has evolved the practice in question for the confusion of interloping Cuckoos, a casual theory which gains some little strength from the finding of an egg tucked away under a mass of feathers after the single baby bird had left the nest. If the Robin's own egg may be hidden in this way, may not the intruding egg of a Cuckoo be similarly treated? Indeed, may not the thick lining act as a screen from the Cuckoo itself, which, it seems, has a thoughtful habit of removing an egg before adding its own product to a clutch?

It is all very big and very wonderful, this problem constituted by a family of birds which have so far forgotten a fundamental principle of life as to foist their offspring upon other birds. And not the least remarkable point in the puzzle is the attitude of the foster-parents. When this *role* is merely prospective they will repulse the Cuckoos with all the fierceness and clamor of outraged respectability; but once the egg is deposited, be it ever so unlike the product of the owner of the nest, it is accepted, in about seven cases out of ten, with a fatalistic resignation, or, rather, a blind trustfulness. What sense of divination is it, then, that persuades birds in the remaining three cases (not necessarily of a particular species) to either throw out the intruding egg or bury it in a corner of the nest?

There is no instance on record, as far as I am aware, of a young Cuckoo being thrown out or utterly neglected by its foster-parents. They may be puzzled at the enormous mouth and insatiable appetite of this freak progeny, but the parental instinct is sufficiently strong to keep them tending it with a faithful persistence that leaves both birds in a half-starved condition. There are even cases in which neighbors lend their aid. Glancing over the records of many Novembers, I find an instance in which a young Fantailed Cuckoo, well able to fend for itself, was being fed by two White-fronted Bush-Chats, while two others did their dainty best to draw me away by feigning to be wounded. All misguided devotion, sure enough, but scarcely more so than that enjoyed by many human idlers, from well before the day of the original prodigal son up to the present enlightened age.

And so, with many a wanton wile, scarcely subdued by an undercurrent of melancholy born of seasonal culmination and the mystery of life and death, November speeds on her busy way to link the hands of Spring and Summer. Then arises that essentially human plaint, "I wish it were always Spring!" It is not a new cry; old writers have even carried it into an insistence on the unchangeableness of heaven:

> A place where all the year is May,
> Where every bird doth sit and sing
> Continually, as in the Spring.

THE PASSING

"No change at all there," runs another foretaste; "no Winter and Summer; not like the poor comforts here, but a bliss always flourishing."

Poor comforts, in sooth! The largesse of Nature may be of scant interest to those who ever look before and after, at the expense of the present, but for your rational watcher of the earth there is throbbing joy in the coming, and sober content in the going, of all seasons. Closely, reader, and still more closely, mark the pageantry enveloping the broad-bosomed mother of us all; then will you find yourself at one with most of the Jovian moods — not wishing vaguely for what cannot be, not seeking needless consolation in the thought of an unchangeable future, but rather associating yourself with the feelings of the poet and the Gull who basked, in golden weather, beside the blue Pacific:

> We two may be forgiven
> If, having found a heaven on earth,
> We ask an earth in heaven.

Chapter V

WITH CHILDREN IN BIRDLAND

When the bird Day movement first found a place in the schools of Southern Australia, the children sat back and vaguely wondered. They were not quite sure what it was all about. No more so were the teachers. Only a small percentage of the pedagogues had given more than very casual attention to the subject before, and the average teacher is no better (nor worse) than anyone else when faced with the prospect of extra work at the same salary. Thus, the bird-study movement was received with a more or less dignified reserve. There it was, however, backed by a recommendation from the United States, and with it the possibility that, after all, the subject might prove rather pleasant.

On that first Bird Day I met a school party in the bush in central Victoria. The teacher was "leading" in the fashion of a Zoo-visiting father — well in the rear — while the kiddies rambled along with an aimless, noisy heartiness calculated to scare every undomesticated creature in the neighborhood. Presently one bright boy spied the nest of a Yellow-tufted Honeyeater. He yelled gleefully, made a hurried grab, and within the next minute was triumphantly presenting the dainty cradle, with its trio of hapless baby birds, to the accredited leader of the expedition. Alack for the spirit of the primitive savage! The diplomatic rebuke that followed was the first intimation to the Australian boy that bird-*observing* is not necessarily identical with bird-*nesting*.

That was, if my memory serves, well over a decade ago. It is a fair period of probation; a school movement that cannot "find itself" in such space of time has no very definite place awaiting it. But the bird-study movement among Australian children has not taken that long to prove its value. The early experience gave Victoria, New South Wales, and South Australia an indication that they had taken up something that bade fair to be a stimulating force in the life of the child; a subject promising to *ease the strain* (and frequent pain) of primary education; something likely, in a phrase, "to live in making others live."

As the impression deepened into conviction with every individual who took more than the too-common, careless interest in the welfare of the school-child, the cult of Nature-study (using this term for want of a better) gained favor rapidly in circles that were absolutely indifferent, if not actually opposed, to it. Not once, nor twice, but many times, I have listened judiciously to gratuitous "confessions" of regenerate teachers in respect of this "pursuit of triflers." They were all on the same lines — based on the principle of "once I was blind, but now I can see." One Victorian pedagogue of standing told the world at large that a boy or girl whose fraternal interest had been claimed by birds showed a marked increase in brightness, particularly in respect of the English lessons.

For a personal testimony, there remains with me a

strong impression left by a contrast in children of two country schools in the same State, situated respectively in "cow" country and a mining area. No. 1 batch of youngsters, blessed with an open-souled teacher, were as intelligent as any among thousands with whom I have chatted, their smartness at mental arithmetic coinciding, even harmonising, with their cheerfully intelligent knowledge of the birds, the insects, and the flowers of their district. And most of these children were waifs, boarded out by the State to dairy farmers!

In sharp, almost harrowing, contradistinction, the elder children of a mining village but four miles away showed a dulness approaching denseness. Moreover, a distressingly large percentage of them were cross-eyed — due, I thought at random, to a continual peering around the huge heap of tailings which fronted the school. Indeed, the gravel seemed to have eaten into souls; wherefore teachers and children had missed the call of the good green Earth — the invitation to Nature's play-parties, "the sign of the joy of the Lord," as Masefield has it. Only — and here my point strengthens — only the infants of this school were bright, happy, and receptive. Mullock heaps and troubled eyes have no place in Fairyland, and (not to speak irreverently) no need exists there for a Saviour to mix spittle with earth and anoint the eyes of spiritually blind.

All this recalls old Dr. John Brown on the desirability of recreating the natural, healthy interest of the child in the beautiful outdoor things of its birthright. "It is not necessary," wrote he, "that everybody should know everything. Is it not much more to the purpose for every man, when his turn comes, to be able to *do* something? And I say that, other things being equal, a boy who teaches himself natural history, observing everything with a keenness, an intensity, an exactness, is not only a happier boy, but is abler in mind and body for entering upon the great game of life than the pale, nervous bright-eyed, 'interesting' boy who is the miracle of the school, dux for his brief year or two of glory, and, possibly, booby for life."

Then Dr. Brown goes on to refer to a pamphlet on "Ornithology as a Branch of Liberal Education," and speaks of its author, Dr. Adams, as "a man . . . who, at the end of a long life of toil and thought, gave it as his conviction that one of the best helps to true education was to be found in getting the young to teach themselves some one of the natural sciences, of which he singled out ornithology as one of the readiest and most delightful."

Figures relating to children's organisations cannot be made very definite, but it is fair to suggest that at least 200,000 boys and girls have enrolled in the Nature leagues of Victoria, New South Wales, South Australia, and Queensland within the last ten years. Two of these States specialise in Bird Clubs, on lines successfully followed in the United States.

"Ten years ago," writes the director of Nature Study in South Australia, in sending some general observations for the purpose of this sketch, "the shanghai was as much a part of the schoolboy's outfit as was his pencil-case, and, accordingly, bush-birds were strangers to the city. The sight of a Cuckoo, Kestrel, or Blue Wren was something to be recorded in the press. To-day things are different, and we find our native birds turning towards the city in increasing numbers. Each year some fresh arrival, not seen for years, is noted by members of the Ornithological Association. Wild birds are losing their timidity. At some schools they come to be fed by the children at lunch-time. I have seen wild birds catching flies from the hats of children and eating crumbs at their feet."

Bird Day has become an institution in most of the Australian States. South Australia makes it a movable arrangement — that is, allows each district to choose the time of the year that seems best suited to local conditions — but Victoria, New South Wales and Queensland usually fall into line on a day in October, what time our Lady of the Spring has got past her chuckles at the breaking of Winter's sway into a broad smile of serenity, and the nesting of birds is in full swing. On that day practically the whole of the school hours is given up to bird-study, the material for this purpose being largely provided by bird-lovers in the form of articles, stories, verses and pictures in the school magazines of the respective States. In these attractive journals Queensland has specialised in recent years, and the result has been a marked stimulus in the fraternal study of the bird-riches of the great State which constitutes the north-east of Australia. Consider a few gratuitous observations from the children: —

It is recorded that at Kolan South on Bird Day no fewer than 25 nests were found *within the school ground*, which is a protected area.

"I wish to join the league for protecting birds," writes a small girl, "as I love these little animals. I have come from the north of Queensland, and I have seen millions of birds. It is a shame to see these cruel town boys, the way they trap and shoot poor, harmless little birds. This is the way I look: Suppose a pair of birds have two young ones to provide for. Some cruel boys come along and shoot the parents. What will the young ones do? Perish, of course. I have no more to say."

Full of matter is this note, signed by five girls of a town primary school: — "Yesterday one of us, Eileen Rivers, saw a boy after a nest; so she took his bag of books and threw it over a fence. He was bigger than she was, so she thought that was the best thing to do. He was afraid of losing his bag, so got down quickly and went after it, then went home." Resourceful Eileen! One is led to wonder how that misguided boy would have fared had he *not* been bigger than the sympathetic small girl.

From a bush school there is this slightly ambiguous note: — "While reading my School Paper I saw that pupils can become members of the Gould League of Bird-lovers, and as I love the little birdies I wish to become one."

Some children claim to have clean sheets in respect of the birds. "I have never caged birds," writes one girl, "and I never will." "I am ten years old," says a bush boy, "and I have not injured a bird yet." 'Twas not always thus. The present writer once confessed, at a meeting of the Melbourne Bird Observers' Club, that his early interest in birds was accompanied by a shanghai; whereat a well-known naturalist burst forth, "Good Lord, man, that's the way we *all* began!"

On the same principle, a bunch of bush-boys candidly admit having often robbed birds' nests, but declare that the cruelty of the practice is now apparent to them. Better still, there is recognition of the miserable lot of the average cage-bird. "I like," writes one lad, making a smart distinction, "to hear birds *singing* in the trees rather than *crying* in a cage."

A pretty story told by a small girl relates to the succoring of a quail. She found the bird, in the wake of a mowing machine, cruelly torn, being minus half of one wing and half of one leg. Taking the hapless creature home, she tended it carefully, and it lived with her for two years before falling a victim to a cat. Again, a warm child-heart comes into evidence in this note: — "A dying baby 'Red-beak' fell out of its nest at my feet. I took it inside and gave it something to eat. Then I climbed the tree and put it back. It seemed all right."

I like also those quaint little generalising notes in which wells out the joy of the Australian child in its land of sunshine and song. For instance: "The human eye cannot imagine how lovely the hills looked." A rapture, this, akin to the midsummer maze of Nick Bottom: "The eye of man hath not heard, the ear of man hath not seen, man's hand is not able to taste, his tongue to conceive, nor his heart to report, what my dream was."

The pleasures of Bird Day rambling — how is a bewitched small girl to set them down? "If I am asked which day is the best in the year I shall say *to-day*." Victor Daley, you remember, gave token of similar enthusiasm:

>When the days that have passed arise
>   On the Day that is to be,
>   When souls leave the land and sea
>And graves under many skies,
>   This day I shall know and say:
>   "I was alive that day!"

"As we walked along a bracken path," runs another record, "I saw what I thought were gleams of the sun, but when we reached the spot I saw what I had mistaken for the sun was wattle." The aesthetic value of that experience was probably not lessened by what followed: "I picked some of the fragrant blossom for the gratification of my nose!"

Rambling observations of quaint expressiveness crowd the recollection. Here are just a few (the italics being mine): — "The Magpie is black plumaged, with white feathers." "The short eggs of the Soldier-Bird will hatch hens and the long ones will hatch roosters." (Behind that forthright information is a fairly general belief.) "The Friar-Bird is a sole inhabitant of Australia." "The nest" (of a water-bird, evidently a creature of millinery tastes) "is *trimmed* with pebbles." "*Haunts* of wild-fowl were hidden in the grass." "In one Bower-Bird's playground we found numerous brightly colored *rages*." The final "e" was probably accidental, but a brightly colored rage is too interesting to be overlooked. "Water-birds fly as though they were reaching out for something to eat in the air." "Some birds" (says a girl of tender years) "take more care of their young than some people do of their children."

Among all these and many kindred novelties, however, memory picks its way to the plaint of a country child who was not allowed pets: "My father says he will keep nothing about the place which is not of use." Poor child! And poorer father! It may be that compulsory devotion to the dairy soured and separated his boyish will from the wind's will, but that seems no good reason why he should carry on the process of spiritual starvation.

Mateship with birds does not stale; but, should the early freshness and wonder dim with yourself, there is a call to help retain it for others. I remember a class of quite small Australians being much exercised in considering whether the ubiquitous "little bird" really deserved its world-wide reputation for sagacity and tale-telling. Presently arose the Sceptic. From the vantage points of a back seat, seven years, and Acquired Knowledge, he laughed the idea to scorn.

WITH CHILDREN IN BIRDLAND

"M' mother," he announced, "said a little bird told her somethin' about me, and I found out afterwards *it was me bruvver!*"

The situation was serious. Tactful handling seemed essential.

"How do you know," I asked, borrowing a trifle of the knowing air, "that the little bird didn't tell your brother?"

In sooth, a brilliant idea at a venture. "Ooooh!" said a sobered scorner, and subsided limply. The thoughts of youth were long, long thoughts that afternoon.

Well, then, if we (or some of us; chiefly our mothers) are able to interpret the message of a small bird, shall not the bird also understand us? Why, certainly:

On the perfect afternoon of a Queensland winter's day I lay on the bank of a small creek near Gympie, with eyes intently fixed on a tree above, into which a yellow bird had disappeared. Presently strolled into the picture three small girls. It was well to be fraternal, but that bird ought not to pass unidentified; and so, without taking my eyes off the tree, I spoke to the children. They stopped at the voice, recognised its owner, and noted the gaze turned aloft. A bit of Fairyland arithmetic followed quickly, and then a pleasantly awed voice spoke. "Ooo-h!" said the girl, *"he's talking to the birds.* Let's go 'way!"

Southward and backward now to days in the bird-realms of southern Australia. There was a time when, in journalistic "between-whiles," I led forth children of another day, those whom the benevolent parson dubs "young people" — prospective teachers, to be precise. Those were live hours.

A girl of fifteen years grew reflective as she watched a group of Babblers vigorously tossing bits of bark about in their insect-hunting. Then a recollection stirred. "My!"

she said, "that's just the way I fling my books away when I'm wild!" (A brightly colored rage, surely!)

Again, a flock of English Starlings swept down the horizon on the wings of a high wind. "Gee!" exclaimed an envious boy. "They can travel at a pace!"

"It's all very well for them," came smartly from a curly haired girl. "They haven't any hats to hold on!"

There was an occasion when a sudden shower forced us to shelter in (of all places!) the clerics' robe-room at a bush cemetery. And there those joyous juniors fired off every bit of scientific bird-nomenclature they knew — all because a wide-eyed old sexton and a wide-mouthed young stone-mason sat agape on the door-step!

Scientific terms attract but few children. One bush lad of twelve years or so, however, developed into a regular technical treatise. In the course of a flower-chat he exhibited huge satisfaction in rolling off his tongue remarks on *Diuris maculata*, *Diuris punctata*, *Glossodia major*, and other orchids. The same boy insisted on carrying my camera later — and dropped it with a breaking thud to ejaculate, in a tone telling plainly he was glad he had been born: "Hullo! *Bulbine bulbosa!*"

Latin terms are very necessary in zoology, but not in the schoolroom or for general use. I call to mind the fund of amusement we derived on a Murray River bird excursion by enlightening an original cook's mate on the scientific names of many of the birds met with. He seemed to derive a vast amount of entertainment therefrom, and was especially struck with the discovery that "the littlest birds get the biggest names!" That ribald Australian was not so eminently patriotic as John Burroughs' guide of the Maine woods, who gained unlimited satisfaction from hearing his common birds and plants honored with good, mouth-filling titles.

WITH CHILDREN IN BIRDLAND

Dwelling on this matter of nomenclature, Burroughs remarks that he once found a little water-creature new to him, and invited a learned man's opinion as to its identity. The reply was comprehensive, the find being fixed as "a species of *phyllopodus* crustacea known as *Eubranchipus vernalis.*" "This title," observes Burroughs, "conferred a new dignity on my fish, but, when the learned man added that it was familiarly called the fairy shrimp, I felt a deeper pleasure."

On the same principle, we want more easy, graceful vernaculars for our Australian birds, names as simple and musical as *Bobolink, Chickadee,* and *Whipoorwill,* before we can hope to have our poets put forward a production to rival the excellent American anthology, "Through the Year with Birds and Poets." Many of the old, heavy names have disappeared, thanks be, but there are others in the way still. Black-faced Cuckoo-Shrike, White-shouldered Caterpillar-eater, Yellow-bellied Shrike-Tit — names such as these (all belonging to common birds) would surely be too much for even Mr. Bernard O'Dowd to weave into a song!

Consider a lyric poet's view on the point:

"Undoubtedly there is," runs a letter from Roderic Quinn, "a vast virgin field for the inspiration of Australian poets to be found in the bird-life of our country. If the lovers of our little friends of the forest and scrub could only succeed in releasing the birds of Australia from the ugly, unmusical, and borrowed names early bestowed upon them, it would do much to even down the poetic path. I suppose, however, we shall have to wait many years ere this is done, and until the sweetening effect of human association gives to our bird-life an irresistible poetic appeal."

That is quite correct, excepting, perhaps, the suggestion regarding the necessity for many years of waiting. Personally, I have tried frequently, chiefly through the medium of daily, weekly, and monthly publications in three States, to evolve euphony in Australian woodland names, and the meagre results of the search have not altered the opinion that much may be done as soon as children, poets, and bush naturalists *join the birds themselves* in the pleasant effort.

Meanwhile, it is well to remember that the sweetening effect operates just as potently the other way about; wherefore, the wise teacher leaves any preaching to the birds — keeps out the officious showman element in favor of the human interest. "On Bird Day," wrote a small girl in an essay that came my way in Victoria, "we had fine speeches given us by *human beings.*" That delightful touch may have been perpetrated unconsciously, but the distinction drawn between the funereal teacher and the breezy, bird-loving visitor (a cleric, if you please!) was too neat and instructive to be lost. As a Solomon Island "boy" remarked to a roving friend of mine, referring to a missionary, "Oh, him 'nother kind. No' all same wite man!"

To the young, Nature is a joyous playground, and the successful teacher is surely he or she who goes closest to using this playground as a means of implanting in the heart of the child, Agassiz-like, the seeds of perennial youth, and the ability to find ports and happy havens in most places that the eye of heaven visits.

PART II

BIOGRAPHIES OF BIRDLAND

Chapter VI

THE IDYLL OF THE BLOSSOM-BIRDS

Life is a gladsome proposition with the Blossom-Birds of Australia. Of course, no one would be well advised to say that such is not the case with every other unit of our winged wanderers, from the Jovian Eagle to the frivolous Fantail. But (speaking of families) the nectar-lovers, above all others, give evidence of keen appreciation of the right strong joy of living. They are, so to speak, the Bohemians of the bird world.

In the next breath, however, it has to be conceded that the birds are of considerably greater value to their country, and vastly more attached to their families, than very many members of the human genus who figure in that quaint category. Moreover, they are a good deal better dressed. Some of Australia's Honeyeaters, in fact, are among the most beautiful birds in the world, and the "tribe" generally possesses almost as many vivid colors as a brilliant sunset. (Yet it does not follow that the apparel oft betrays the bird; there is a wonderful degree of harmony between the habitat and plumage of even the brightest of Nature's children.)

It were unwise — and that not necessarily for the sake of the bird — to pursue our parallel between Bohemian man and Bohemian bird on certain other points. It is true enough that Honeyeaters have been found helplessly intoxicated beneath certain flowering trees, but we may safely put those odd cases down to pure inadvertence — as distinct from sottishness. Nor would we be unduly gratified by considering which of the two high-living roysterers (man or bird) gives the less thought to the morrow. Suffice it to note that the happiness of the Honey-eater is literally a joy for ever; "the contagion of the world's slow stain" is an unknown quantity in the life of a free, wild bird.

As a matter of course, life in Australian bush or more settled areas would be a good deal less joyous without the presence of representatives of the Honey-Birds. They are the most characteristic, not to say novel, and largest family of birds found in this continent. All but one of more than 250 species known to the world are found in the Australian region (from Wallace's line to the Sandwich Islands and New Zealand), and something like 90 of these belong to Australia itself. A happy arrangement, this! It is as though Mother Nature, having decided upon eucalypts and other myrtaceous trees as the chief feature of Australian vegetation, evolved these birds as kin-spirits of the blossoms, giving them long bills with which to explore the flowers, and sensitive, brush-tipped tongues with which to sweep up the nectar, not to speak of a score of cheery attributes calculated to win the approval of the august lords of creation. And the Honey-Birds have done their part with right good will.

Indeed, they have gone one better than their original duty of fertilising flowers, by taking to an intermittent

THE IDYLL OF THE BLOSSOM-BIRDS 75

diet of insects. It may be, as Mr. Pycraft (of the British Museum) suggests, that the Honeyeaters began to vary their "menu" with insects through brushing these up with the pollen in flowers; but, at all events, the fact now is that most of our "sweet-tongued" birds go out of their way to catch insects. This point may readily be proved by anyone who cares to watch the Wattle-bird or any of the common smaller Honeyeaters at work, and it is a point that should be placed to the birds' credit by those fruit-farmers who, seeing the long-billed birds in the orchard, are apt to overlook the service they render, and do something rash with a gun.

It has been suggested earlier that the Honey-eaters, more than all others of our avian revellers, are of vitality compact. They have to be. Flowers may come, but flowers quickly go, and it is necessary that their kin-spirits "make the most of what they have to spend," and, moreover, keep moving about as the seasonal flow of the blossoms requires. The majority of the Honey-Birds, therefore, are born travellers, or, to be more precise, confirmed wanderers — happy-go-lucky vagrants, who follow the fluctuations of the flowers from district to district, or State to State, according to the necessities of the day and the hour.

That, of course, is in the cooler months. Later on the birds must needs remain loyal to a particular locality for nesting purposes; but from March to July they live in actual fact the idyllic life of which "my dainty Ariel" sang:

> Merrily, merrily, shall I live now,
> Under the blossom that hangs on the bough.

Here, again, is evidence of the thoughtfulness of Mother Nature for her children. For is it not in the Winter months that the great body of the eucalypts are in blossom? And when the supply of nectar eases off in the Spring-time are there not many insects upon which baby Honey-Birds may be fed? For answer, watch the insectivorous Cuckoos in search of hosts for their intruding eggs — how they patronise the Honey-Birds.

The case of the Lorikeets, those dapper, nectar-loving Parrots which are more or less familiar to dwellers in every Australian State, is somewhat different. Their period of nesting, in Southern Australia, at all events, is later than that of the Honey-birds proper, and it would appear that Spring sometimes means to them a shortage of food. At that period I have seen Lorikeets feasting with obvious satisfaction among the introduced poplars and pine trees (*P. insignus*). By the same token, they are the fastest and most confirmed travellers of all the Honey-Birds. As a general rule the main body of the Musk, Little, and Purple-crowned Lorikeets winter together in the big gums of the Southern States, but in an unfavorable season the first two species may call on Queensland, leaving only their congener with the purple crown to remain faithful to the South. Conversely, the gorgeous Rainbow (Blue Mountain) Lorikeet often leaves Queensland far behind, whereas its companion of the North, the Scaly-breasted Lorikeet, is never persuaded to experiment lower than New South Wales.

At all times the Rainbow and Scaly-breasted Lorikeets are common in Queensland, where, sadly enough, the pretty birds — commonly known as "Bluies" and "Greenies" — are trapped in thousands and condemned to screech unhappy lives away in cages. The astonishing thing is that people who can see the birds at their best, revelling in the air or the trees, on almost any day, should

THE IDYLL OF THE BLOSSOM-BIRDS 77

find satisfaction in having them as miserable captives.

The queerest vagrant of all the blossom-birds is not a Lorikeet, however, but the Regent Honeyeater — the "Coachie" of Victorian boys and the "Embroidered Honeyeater" of a forgotten artist of 1811 — a medium-sized nectar-lover, whose beautiful plumage of yellow and black gives it rank among the beauty-birds of Australia, and, accordingly, the world. The species is rare in Queensland, occasionally seen in New South Wales, and more numerous in Victoria and South Australia. I knew the bird well in Victoria, where it usually was to be seen, and its melodious, flute-like calls heard, the whole year through. There were odd times, however, when this gorgeous reveller seemed to take into its collective head the idea that a change of air would do it good; whereupon a district flowing with insects and honey would know the bird no more for, perhaps, several years. For instance, every one of the numerous Regent Honeyeaters which had long been about central Victoria departed for fresh forests in the early Summer of 1913, and never a sign of them was to be seen during the whole of the following year — this despite the fact that all other species of honey-eating birds were as numerous as ever. Even a pair that had nested in the same tree for three successive years failed to return. What took those birds away? And where did they spend the intervening period? These are problems still unsolved. I know only that there was keen pleasure (and some relief) in more than one breast when the mellifluous "clink-clank" of the prodigals echoed again about their old haunts in August of 1915.

It is a happy coincidence, and, incidentally, a notable corrective of the old "songless bright birds" impression, that the Honeyeater which shares with its Regent relative the family honors for beauty, is also the possessor of a most musical voice. This is the fiery little sprite known to ornithologists as *Myzomela sanguineolenta*, the Scarlet Honeyeater, and to less fastidious folk as the "Blood-Bird." It is a member of a genus which numbers over 50 species, half a dozen of which enliven the Australian landscape. Essentially a denizen of the richly vegetated areas of the east coast, the Scarlet Honey-Bird is seldom seen in Victoria, is occasionally numerous about Sydney in the Summer-time, and is quite familiar to coastal Queenslanders in the Spring. Individuals may be also heard about Southern Queensland in the cooler months, but the main body of the birds are then among the flowering trees and shrubs in the far North.

The tea-tree (*Melaleuca*) and bottle-brush (*Callistemon*) are its favorite hosts. Wherever and whenever these flowers are out, there and then may the silvery, bell-like tinkle of the Blood-Bird be heard, and, if the observer be keen of eye, the scarlet form seen flashing about the tops of the tall trees. There are times when it comes into low trees. One of the prettiest avian sights I have ever seen was constituted by a small host of these birds flitting rapidly among whitish flowers which bedecked a small tea-tree glade at East Brisbane.

How long a period is covered in the attaining of the scarlet livery of the Blood-Bird is an open question. An acquaintance who has had a specimen in an aviary for five years assures me that it is only now beginning to color. Unnatural diet, however, is probably the cause of this and other records of lengthy periods taken by captive birds to come to their own; and we are not holding, as it were, a mirror up to Nature in supposing such protracted adolescence to be natural. Is the life of a wild bird so free from hazards that it can afford to spend five or seven years of precious youth in convent garb? I trow not. And yet, on the other hand, it is obvious that the attainment of bright plumage adds to these risks, bringing also to most of its owners a sense of responsibility and discretion.

Not so with the careless little cardinal of the blossoms. I see more of the flashing scarlet forms than those of sober dress, and, what is more, hear from the female only a modest "Chip-chip-chip!" a vocal performance very different from the musical rhapsody of her gay little lord. What a life is his! — an idyll in itself! Beauty of voice, form, plumage, and movements, all are embodied in this brilliant bird; and, moreover, his love of tropical nectar ensures him a permanent place in the sun — the "fetterless, idyllic round of enchanted days" which O. Henry chanced upon in Latin America.

"Careless little cardinal" I have called the head of this branch of the Honeyeater House. But in the next breath it has to be said that the gay sprite is not so careless as he might be, and as I once took him to be. In September of 1920 a pair of Scarlet Honey-Birds were found in charge of a fragile, pendulous cradle placed in a cypress pine-tree hard by Redland Bay, South Queensland. At the first visit the graceful mother-bird flew off, disclosing a tiny young one and a single egg. On the following morning the little grey mother was again in charge, and, as she displayed no disposition to move, I stirred her gently with a stick. Straight to the ground she went then, and fluttered along with every indication of being badly "winged."

What a surprise this was! Somehow, I had achieved the impression that nothing of the kind entered into the powers of a small inhabitant of the tree-tops. Was it the use of the stick that prompted this pretty act, which, be it noted, was not "put over" on the first occasion? His Grace the Cardinal had stayed severely away on my first visit, but this time he came upon the scene, and, with an air of profound importance, not to say responsibility, he accompanied the timid wee mother on each of the brief stages by which she hesitatingly approached the nest. The last stage of all — the hop to the nest — she had to negotiate alone. This accomplished, Red-Bird gave a prideful little tinkle and flashed away.

"Sir," said I, with Pickwickian fervor, "you are a humbug!"

Behold, though, how easy it is to misjudge even a tiny bird! In very few minutes the brilliant Cardinal, he whom I had rashly assessed as a shirker, was back at the nest feeding his devoted mate. He left me quite penitent, and when, a few minutes later, he came again and fed the two babes, I took back all I had ever said or thought of the Scarlet Honey-Bird as an unpractical dandy.

To any bird-lover with knowledge of both South and North of the big State of the Tropics, memories of the Cardinal Honeyeater must ever be associated with that other little beauty of the blossoms, the Sun-Bird. Sired by a sunbeam, born of a flower — that is Mr. E. J. Banfield's impression of this dainty creature that flits day-long about the scarlet hibiscus flowers on Dunk Island. My first sight, there, of the vivid little male bird will remain long, as long, perhaps, as the initial glimpse of the radiant creature remained with the pioneering Macgillivray, the naturalist of H.M.S. *Rattlesnake*, after he met the Sun-Bird back in the 'forties.

With its olive-green back, blue bib, and yellow breast sheening in the tropic sun, my bird sat in proximity to a flowering bush of hibiscus. Occasionally he broke into

THE IDYLL OF THE BLOSSOM-BIRDS 81

a semi-plaintive, semi-placid "Purr-r-r," rather like the call of an English Goldfinch in anxious mood. Again, there would well from the bright breast a spirited chatter, not unlike that uttered by its kin-spirit, the Spine-billed Honeyeater. That was one of the few occasions upon which I ever saw a Sun-Bird resting. Embodiment of the tropic Summer, without any of its langour, he (or she) has all the high vitality of the famous Humming-Bird, and, added to the ability to hover in front of a flower, the Sun-Bird has a spasmodic habit of rushing pell-mell away on an aimless errand that concerns no one, not even itself.

Marking all this, would you not expect such a bird, when nesting-duties interrupted its flickering and frisking, to build and brood somewhere in the tree-tops — to approach the alleged habit of the Birds of Paradise, which once were fabled to carry their eggs on their backs? But, note the anomaly. For all its freedom, all its airy aloofness, all its self-centred revelry in sunshine and flowers, the Sun-Bird chooses, in seven cases out of ten, to suspend its domed cradle from man-made buildings. Is this done in spite of or because of its elfin life? The answer would seem to be this: The Sun-Bird, like the Swift, is too electrically busy to have time to become afraid of slow-moving humanity. Angels may weep at some of the tricks performed in the name of human authority, but to this independent little Ariel (an Ariel in all but servitude) Man simply does not matter. By the same token, however, if earth-cumbered folk choose to provide flowers upon which birds may feed, and verandahs under which nests may conveniently be hung, it is decreed that no Sun-Bird need carry its pride to the point of being standoffish.

Reference having been made, in passing, to the lyric music offered by several of the blossom-birds, it is necessary now to tender a slight qualification on behalf of the two families (Honey-Birds and Sun-Birds) generally. Few of them have what may be termed straightforward singing powers. But what of that? Their voiceful persuasiveness rests in the heartiness and good cheer contained in the chortles, lilts, shouts, and widely varied calls generally, from the irresponsible revelry of the Wattle and Friar Birds, past the companionable chatter of the curious Fasciated Honeyeater of Queensland, down to the altogether blithesome notes of the small White-naped Honeyeater (Black-cap) of the South. Truly, Dennis, "it's the sunshine of the country, caught and turned to bonzer notes" that the cheery Honey-Birds pass on to whomsoever will listen. It is almost literally true also that there is sunshine in the movements of these graceful birds, whether they be those of the Lorikeets screeching overhead in the crisp air of a Winter morning, the beauteous Spine-bill fluttering, Humming-Bird-like, before the bells of the heath, or the movements of a dozen other strong-clawed species — comrades of the dancing leaves — performing acrobatic feats among the blossoms.

Here, too, I think of the *real* dancing of a blithe companion of my boyhood days, the Yellow-tufted Honeyeater. This pretty study in brown, yellow, and black is one of many smaller members of the family which remain more or less constant to the same locality the year through — that is, of course, when conditions have not become sufficiently unfavorable, as has been the case in the Brisbane district, to drive them out altogether. And in the Autumn, when there is none of the cares of the nesting season to distract, the exuberant spirits of the Yellow-tufts frequently carry them through regular quadrilles.

THE IDYLL OF THE BLOSSOM-BIRDS 83

I recall one such performance, which was presented by about 30 birds. It was most engaging. With wings drooped and tails raised, the graceful creatures bowed, advanced, retired, hopped around, and, amid much excited chattering, went through a highly spectacular display before rising simultaneously and flying to another spot to repeat the antics. Do the birds deliberately arrange these play-parties? It seems to me that, in a semi-capricious way, they do, even though we have nothing to show how such arrangements are effected.

Quite apart from that consideration, however, there can be no doubt as to the purpose of the Yellow-tuft in its even more extraordinary antics when danger appears to menace its babies. No bird of my acquaintance is able to practise the broken-wing ruse with greater skill than this Honeyeater. Uttering the most pitiful of cries, it will go fluttering along over the roughest ground, falling into depressions and scrambling out again, now beating its little wings on the earth, now making them to quiver tremulously in the air — but all the time keeping a wary eye open to see whether the intruder is being satisfactorily deceived into following away from that precious nest. There never yet was a dog (and seldom a boy) able to resist such invitation, and never did dog (or boy) look more sheepish than when the "wounded" bird gave an amiable "cheep," shook its little tail, and flew quickly to the shelter of a branch. Sometimes the birds make this nest precaution a matter of common concern. I remember an occasion when the fright of some young Yellow-tufts caused all their relatives in the vicinity to rally to the defence. The whole company raised a terrific hue and cry, and at least *three* of the birds gave superb exhibitions of feigning to be disabled. This community of interest, by the way, is much in evidence among those common Honeyeaters known as Soldier-Birds, and also among the beautiful Regent Honeyeaters.

As a matter of course, just as there are wide variations in the size and color of Australia's Honey-Birds, so are there wide variations, if not broad distinctions, in the architecture of their nests. These range from frail platforms of sticks, such as are run together by the Wattle-Birds, past the fibrous home of the Soldier-Bird and the bark nest of the Regent Honeyeater, to the dainty, swinging cradles of the smaller species. It is all quite natural. Apart from the fact that their bills are too heavy for the construction of artistic nests, the larger Honeyeaters seem much more careless than such, for instance, as the Black-cap and the Yellow-tuft. Is it a matter of evolution following upon necessity for protection? Probably the ancestor of the "tribe" was content with a scanty platform of sticks such as is still favored by the Wattle-Bird. Then, as the ages rolled on and the family branches spread wider and wider, aesthetic sense developed along with the need for the smaller birds to use measures to protect their eggs and young in the struggle for existence.

It may be that not very long ago our friend the Yellow-tuft built its home in the fork of a tree. At present the bird "balances" curiously between the suspended and the supported types of nest. As a general rule its nest is pensile, situated in bushes from two to ten feet high. Very often, however, it chooses a situation where the hanging cradle will have support from below. Sometimes it nests right down in tussocks, or adopts protective coloration by nesting in dry bushes; and, again, I have known an instance in which a curious fancy, rather than circumstance, apparently threw a pair of these birds back

THE IDYLL OF THE BLOSSOM-BIRDS

to the habits of their ancestors. There were quite a lot of suitable trees in the neighborhood, but these Yellow-tufts chose deliberately to set aside the "rule" of the species in respect of fibres and a swaying branch, and reverted to bark-threads and the leaf-crowned fork of a sapling.

Such a definite departure as that is rare indeed, but there are numbers of instances wherein the Yellow-tufts have been known to build what a small boy called "fancy" nests. This species is, in fact, the Autolycus of the bird world — confirmed "snapper-up of unconsidered trifles" — and the nests are often found to be reinforced with all manner of small, soft articles, from bits of sack-bag to the dried heads of flowers. One nest I knew was almost wholly built of feathers from a neighboring farmyard; strangely enough, it survived the breezes.

It is not often, however, that the birds take liberties with the weather. I recall that in the Spring of 1913, the Yellow-tufted and Fuscous Honeyeaters built earlier than usual in some districts, the reason becoming apparent in the high winds of September. Moreover, they gave additional strength to their nests, the Yellow-tufts by abandoning the swaying in favor of the supported type, and the so-called "Linnets" by introducing wool among the fibres of their tree-top, breeze-blown homes. And when wool is woven as tightly round branchlets and bushes as these little builders can manage, you may be quite sure that the branch will break before the cradle will fall. But this novel nesting material proved unexpectedly dangerous to the young birds. There came under my notice no less than three cases in which baby Honeyeaters, apparently when opening their beaks to be fed, got their fine brush tongues entangled in the wool strands, and (the parents being unable to help them) perished miserably. In one instance the baby bird had sufficient of life to warrant an effort at succor. It was too late. Removal of the wool and application of moisture showed the top of the tongue to be quite dry and the lower part pitifully swollen.

Those who know our Honey-Birds at all well will remember that, in addition to what has been emphasised regarding their uniform sprightliness, all the species carry a broad general relationship in build — slim body, long bill, and strong feet. (The claws of a young Gippsland Bell-Miner, by the way, I found to be, in a comparative sense, the strongest of any baby bird I had ever met.) It has been pointed out also that the nests and nesting habits of the 90 species are various rather than distinct. And now it remains to be noted that the general evidence of relationship is carried through to the eggs. May it not be that the honey diet has something to do with the delicate pink "back-ground" coloring of the eggs of this characteristic family? But then the Lorikeets, which are equally fond of nectar, lay pure white eggs in the hollows of trees; the sweet-voiced Silver-eyes, which may often be seen in city gardens along with the pretty, yellow-winged, White-bearded Honeyeater, lay pure blue eggs; and the Wood-Swallows, three species of which I have known capriciously to feast heartily on the nectar of silky oak and eucalypt, also produce eggs very differently colored to those of the Honey-Birds proper.

Heigho! shall we ever have an end to these pretty puzzles? And what a world of prose this would be if we had!

Chapter VII

THE ARISTOCRACY OF THE CREST

It might reasonably be said that all crested birds are notable figures among their kind. Certainly, there is a perfectly definite distinction attached to species which possess this lively ornament.

The comparative rarity of the crest is the first factor in this consideration, but a more potent one is the fact that it gives to its owner a sprightliness, and, indeed, a dignity, which many birds of more brilliant plumage cannot display. I would not go so far as to compare the crest with a woman's "crown of glory." Obviously, the former is a much less essential decoration — that is to say, the bird could better afford to lose its crest than the woman could to part with her hair — so much so that one wonders what purpose brought it into being. Howbeit, the fact remains that the distinction of the crest is not lost upon its owner. I have never yet seen a crested bird, young or old, that did not display some indication of the possession of a belief that he (or she) was one of Nature's anointed.

You see these qualities of sprightliness and conscious dignity exhibited by the Cockatoos, and particularly by the pink (Cockalerina) species, which verily appears to have an assured knowledge of the fact that it possesses the most beautiful crest of any bird in Australia. Alas for its owner, these variegated feathers are, like the golden crowns of the old-world Hoopoes, all too fatal in their beauty! The Pink Cockatoo is rapidly becoming one of the rarest of its race. Happily, such is not the case with the regal Sulphur-crested Cockatoo, which (despite the demand which exists for it as a household pet) is still to be found in considerable numbers in most free spaces of the land. Nor does it appear that the several other crested Cockatoos are in any immediate danger. Even the lovable little Cockatoo Parrot seems to be holding its own fairly well. A gentle, affectionate creature this, but withal possessing something of that perky dignity which is, so to say, the sign of the crest.

Apart from the Cockatoo tribe, there are very few of our larger birds decorated with crests; and among the passerine (perching) birds only four genera are so distinctive — that is, excluding the Helmeted Honeyeater, whose "helmet" is more a tuft than a crest, and one or two Flycatchers, whose curious little bunches of head-feathers are really only half-crests and seldom remarked on. The four others referred to are the Whip-Birds (three species), Shrike-Tits (three species), Wedge-bill (one species), and Bell-Bird (one species), each of which is a remarkable, purely Australian genus. They differ a good deal in choice of habitat, but have, on the whole, more points in affinity than otherwise. The Wedge-bill, a hermit-like recluse of the interior, may be regarded as an inland analogy of the shy Whip-Bird of the coastal fastnesses, and the Shrike-Tit has often struck me as being an arboreal edition of *Oreoica*, the Bell-Bird.

THE ARISTOCRACY OF THE CREST

Possibly this latter pair of birds are only distantly related on a scientific basis, but, as there is no other genus between them, they are always put side by side in the school of ornithology. Maybe they were, "in the dark backward and abysm of time," brothers akin, but, as the ages rolled on, one section of the family grew as fond of the tree-tops as the other was of the ground, until, finally, there evolved some very distinct differences between them. And yet they have much in common. There is the crest to begin with, together with a somewhat similar scheme of color markings about the head and throat; then, they both have the quality of ventriloquism in their notes, and, what is more marked, a very similar "charring" chatter when excited. Again, they are both endemic birds, and are almost always to be noted in the same class of country, the lightly timbered eucalypt forests, though the Shrike-Tit is not averse to dwelling in the thick, damp jungles of the east coast if there be sufficient gum trees therein for it to forage and gambol among.

The Crested Bell-Bird (*Oreoica cristata*) must not be confused with the silver-voiced Bell-Miner (*Manorhina melanophrys*) of Kendall's poem. The latter bird is a Honeyeater, which does not venture away from the thickly vegetated areas of the eastern seaboard, and, moreover, it is as communistic as the Bell-Bird is solitary. A choir of Bell-Miners, each one supplying an individual "Tink," is required to create the musical chiming immortalised by the poet, whereas our crested friend of the interior gives a beautiful solo performance. As a natural corollary to this fact, his is the more varied music; his voice is fuller, rounder, more golden in tone, and echoes "with ring and with ripple" of a melodious quality that cannot be equalled by the silver-bell-voiced bird of the gullies.

As a boy in Victoria I knew the Crested Bell-Bird very well. Its mellow chiming was one of the most characteristic sounds of our bush, and, following the commonsense boyish practice of allowing a bird to choose its own name, we knew it as "Dick the Devil" and "Whack-to-the-rottle." The latter title was rather uncouth, but, listening again in fancy to the peculiar, liquid run of notes, it seems to me that the juvenile ear rendered them as near to human speech as was possible in the words, "Dick, Dick-Dick, the Devil" — the whole phrase to be taken leisurely, with, on the last syllable, a liquid drop as that of a small stone splashing into a pool or a soft "clicking" of a human tongue.

Never by any chance is the Bell-Bird persuaded to ring its ventriloquial chimes in a hurried manner. But even the most sober-minded of Nature's children must make some concession to the sweet o' the year; and so, when the fire of Spring is in the air, the Bell-Bird frequently repeats his golden roundelay in a higher key, introduces wavering variations into the first portion of the melody, and adds one or two other syllables before dropping into the final pearl of sound — probably the most melodious single note in the whole range of our bird-music. It is not often that the fairy ringer can be seen at this performance, and even your keenest listener may be left in the bewilderment of a Ferdinand crying on a magic isle: "Where should this music be? In the air or the earth?"

For one thing, the unsubstantial nature of the rolling melody renders it difficult to locate the producer, and, what is more, he (I have not known the female to "ring" at all) has no desire whatever to be more than an unbodied joy to visitors. The slightest sign of intrusion into his bush recess while the spiritual chiming is in progress is

THE ARISTOCRACY OF THE CREST

sufficient to silence the pure-voiced bird.

That is not to say, however, that the bird is not often seen. A species which dwells in every State must needs be fairly well known by sight, and in the case of the Bell-Bird this factor is stressed by a confirmed habit which the bird follows of hopping along roadways. Primarily, this practice has arisen from the bird's discovery that there are many insects to be found about highways, but I suspect also that the sprightly creature has, in contradistinction to his attitude when chiming in a bush recess, a very human desire to see and to be seen. Of course, you never hear song of any kind from a Bell-Bird under these conditions. Occasionally he will emit a chattering "Charr-charr-charr," and, if the observer be lucky, it is possible that he may hear a melodious "Clup-clup" as one bird calls the others on their way.

Nor is the nest of the Bell-Bird very difficult of discovery. Being essentially a stay-at-home species, it builds its domicile and brings up its young in the locality in which it has spent all the rest of the year. The nest itself is a neatly built structure of bark, usually placed upon a bushy stump at a height of three or four feet. Sometimes it is situated quite close to the ground, and, in odd cases, it consists simply of a few pieces of pliable bark woven into the top of the deserted home of a Babbler. I found many nests of the Bell-Bird in the Spring of 1914, some influence in that dry season evidently encouraging this particular species to breed more readily than usual. In October (the favorite month for nesting) I tried many times to photograph the shapely male bird at the nest. But the brilliant eyes — eyes as luminous as those of the Satin Bower-Bird — were too watchful, and never once was I able to persuade their owners to sit for pictures.

The most curious experience I ever had with the peculiar nestlings of the Bell-Bird — babies with all the individuality of their parents — concerned a pair which came into being during a previous October, in a nest placed upon a bushy stump. They were uncanny little creatures. Though almost fully fledged, the tops of their heads and a patch right down each back were quite free of feathers, and, instead of the usual wide-eyed stare of inquiry the visitor gets from most young birds, the eyes were tightly closed. Both babes, in fact, might have been quite devoid of life; but when I touched them lightly there was a decided change of tactics. The eyes remained closed, but the necks were outstretched, the sprouting feathers on the forehead started, and the heads waved in the air in exactly the threatening way of the tails of processional caterpillars! Meanwhile, unlike most other young birds when disturbed, the queer little creatures made not a sound; nor was there any protest from the parents; they, indeed, showed their confidence in the instinct of their babies by remaining severely away from the nest.

Less than an hour later I stole a march on the young actors — and caught them sitting up complacently, with eyes wide open! But they were closed again immediately, and the uncanny mimicry broke out with renewed vigor. It was altogether an extraordinary instance of the high degree of artistry to which the protective instinct will lead some of Nature's children; and it becomes the more eloquent in the light of a similar record made by Dr. Macgillivray, an ex-president of the Royal Australasian Ornithologists' Union, in connection with the little Black-throated Warbler, of North Queensland. "The young birds," he says, "have four peculiar head-plumes, which they have the power of erecting and quivering vigorously.

THE ARISTOCRACY OF THE CREST

When one looks into a nest these head-plumes are put into motion by the birds, and remind one of a number of caterpillars waving about."

Is there, one wonders, any affinity between this queer behaviour on the part of the Bell-Birds and their equally remarkable habit of stocking their nests with hairy caterpillars? Personally, I have noted only the one instance of the waving process by the young birds, but I have always found the employment of caterpillars in the nest to be constant. You see these larvæ — usually, in my experience, those of various moths, with a sprinkling of the caterpillars of the painted lady butterfly — among the eggs (I have even seen them in nests before any eggs had been laid) and, more often, placed upon the broad rim of the saucer-shaped structure. Sometimes they are sluggishly alive, sometimes they are evidently dead, and sometimes they seem verily to be *petrified*.

Mr. G. F. Hill, in an interesting observation upon the subject, gathered when with the Barclay Expedition of 1911–12, says that he found about one dozen living larvae of a *Spilosoma* moth in a nest of a Bell-Bird containing two eggs, in the Northern Territory. He notes that it is not uncommon to find the larvae of *Spilosoma obliqua* in Bell-Birds' nests in Victoria, but considers it somewhat remarkable to find larvae of a very closely allied species in the nests of Northern Territory birds.

But the whole practice is remarkable; so much so that, in the absence of any definite evidence to indicate that all these hairy caterpillars are stocked for food, the purpose of the custom still remains a puzzle to ornithologists.

And now, all ye who respect a highly capable and dainty bird, consider with some closeness the Bell-Bird's kin-spirit of the crest.

It is not because of any particular rarity that the Crested Tit has not been generally studied. Broadly speaking, the distribution of the group is somewhat akin to that of the three species of Whip-Bird. The most numerous is the yellow-breasted species (*Falcunculus frontatus*) of Queensland, New South Wales, Victoria, and South Australia; a white-breasted form inhabits West Australia; and a third species, known as the Yellow Shrike-Tit, was recently discovered in the far nor'-west. As the names signify, each of the trio has its distinctive coloring, but their relationship is very marked in the matter of size (about that of a Sparrow), general build (particularly the sturdy beak), and sprightly crest, this ornament acting admirably in setting off the pretty plumage.

Each Shrike-Tit's mode of life, too, is much the same. The greater part of their time is spent high up in the trees, either attacking leaf insects, or, with their splendidly strong bills, waging war on the many borers which live beneath the bark. (Only once have I seen one of these birds emulating the Bell-Bird by working on the ground.) When an observer has paid some attention to the Shrike-Tits he can soon locate them in forest country by hearing the sound of hard beaks hammering on trees in much the same way as Wood-peckers in other countries — for which reason the birds were the "Yellowhammers" of our boyhood days. Among the leaves the presence of the hard-working little acrobats is denoted by a crackling sound, resembling that created by a feeding Parrot. Indeed, the beak of this crested enthusiast is just as strong as that of a Rosella.

I was about to say that if the Shrike-Tit is not rare the same cannot be said of the nest. But the nest of any well-distributed bird can hardly be rare; the point is that

THE ARISTOCRACY OF THE CREST

it is extremely well hidden. In those early days I had often wondered where the shapely yellow-and-grey bird with the black-and-white head had its home, but it was not until a few years ago that I gained any definite knowledge regarding what a capable naturalist has called "the find of a lifetime." In the spring of 1912 I noted a pair of Shrike-Tits, with a trio of fledglings, about a certain belt of timber in a bush recess, and in the following year gave that locality close attention. The Shrike-Tits are constant to a favorable area, and, sure enough, the pair in question(?) ceased their happy-go-lucky wanderings in August, and came again about a fossicker's camp which they had been wont to patronise.

For several days in the following month we tried to trace the male bird as he left the camp, but he always flew in what was ultimately found to be a misleading direction. However, he did contribute to the locating of the nest by the utterance of a loud, penetrating, ventriloquial monotone. It was near the end of September when, having watched the black-throated male bird in a particular tree until my neck ached, I was rewarded by seeing the green-throated female arrive, flash up to the topmost fringes of the leaves, and weave away at a small cluster that was assuming cup shape.

What satisfaction it was to have found the precious nest at last! It was a privilege, also, to sit by while the Tits worked, and be supplied with practical evidence against the libel that the male has either no ability or no inclination to assist in nest-building. The female is certainly the leading spirit in the enterprise, but (in addition to supplying the incidental music) the male does his part by bringing a share of the materials, and at times too, varies the practice of passing the fibres to the female by stitching them in himself. Probably only the more delicate worker attends to the weaving and binding of the web-like substance on the outside of the nest. Around the rim this is managed by the bird sitting in the nest that its breast is modelling, and drawing its bill gently upwards — a pretty practice that the Shrike-Tit has in common with the Fly-catchers, and some other birds which build soft, open nests.

In noting these nesting arrangements I had not only the one pair of birds to rely on; for the experience gained with them helped in the locating of several other nests, and by mid-November I had listed seven of these dainty dwellings. None was situated lower than 25 feet, and some few swayed at the tops of trees 50 feet in height. Of necessity, they had to be very deep to ensure the safety of the eggs or young, and even then, when the winds blew, how the cradles did rock! In some instances only the tip of the brooding bird's tail could be seen; indeed, it was a puzzle how she (or he) reached the eggs at all. Almost invariably some of the leaves above the nest were nipped off by the birds. Why? A school class to whom I showed one such nest suggested that it was to lessen the swaying of the branchlets — a logical assumption. At times, too, the vertical branch to which the nest was fastened — it is never suspended — was cleaned of its bark in proximity to the pretty home, for what reason it is not easy to suggest.

But the most remarkable feature in connection with the nests came under notice when a few of the deserted homes were collected at the end of the Spring. Examination of the material proved what I had already suspected by seeing the strong-billed Tits hammering at the *green* bark of trees, that the nests are constructed almost wholly of the yellowish, fibrous bark underlying the rougher exterior of the limbs. This is beaten into fine threads, and bound tightly with the filmy substance from cocoons, and, occasionally, lichen from the old "snake" fences. Of grass little is used — just a dozen or so fine bents at the bottom of the nest.

How curious it is that the Shrike-Tits, for all their finely safe nesting sites, occasionally fail to bring forth a brood! What persuades the birds sometimes to desert a completed nest for days at a stretch is hard to reason out. Nor is it less difficult to account for some of their dwellings having a ragged hole torn in one side. The only thief I ever saw at a Shrike-Tit's nest was an enterprising Honeyeater, which, when I was coming down after vainly endeavoring to peep into a tree-top home, flitted up and stole some of the soft material, leaving me to bear the blame.

Shrike-Tits are usually very suspicious when building, and there is a danger of them deserting a nest if you watch the constructive operations too closely. In one case where this happened, my clumsiness was partly excused by the suspicion that the female did not approve of the site chosen. She took no active part in the nest-construction for a full half-hour after I first located the birds. Meanwhile, the industrious male bird hammered away at the bark of green trees, and wove the material gained into a little tuft that could faintly be spied growing at the tip of a gum 45 feet high. Betimes, he grew anxious, left off the work, and called loudly for that truant mate.

She returned soon afterwards, and operations were resumed conjointly; but when I went to note progress a week later the cluster in the tree-top had not grown, and the birds were missing.

There are odd occasions when Shrike-Tits, a trifle more capricious than their brethren, build at a much lower height than is usual. A Victorian school teacher has told me of a case in which a pair of the birds nested as low as ten feet. Such a departure is surely remarkable, for the greatest factor in ensuring safety of the nest is its inaccessibility. It is not surprising that the pair in question was three times robbed by a cat; they showed more pluck than discretion in sticking to the one spot for so long.

Lowliest of my own Shrike-Tit "finds" was a nest situated at the top of a sapling 17 or 18 feet high, and not far from a busy country road. And thereby hangs a tale, one directly connected with the frontispiece of this book. Such a modest height was low enough to render a photograph possible; accordingly, we brought into operation, early on a bright November morning, what a facetious cyclist was pleased to term a "wireless station." Engrossed in the interests of three lusty babes, the parent birds did not appear to resent the presence of a long-legged apparition with a staring eye that peered in at their treasures; they came back to feed the chicks almost immediately the camera was in position, and the faithful male parent sat on the nest with wings spread wide to keep the hot sun from scorching the callow young, what time his beak was wide open, as though the bird suffered from thirst.

The opportunity for photographing the birds at home was certainly magnificent. Trouble arose, however, through an error of judgment regarding the length of the improvised tripod. That checked operations for the day, for the horse, cart, and other apparatus had to be returned safely before its rightful owner was out of bed. Then, as fate would have it, dawn of the following morning was accompanied by a high wind; the nest swayed dangerously, and photography was quite out of the question.

THE ARISTOCRACY OF THE CREST

And so came a third trip. There would be no mistake this time! The camera was in first-class order, the long "legs" were strapped tightly to the tripod, the wheels of the cart were safely locked, to guard against any unscientific tendencies on the part of our venerable horse, and, above all, the weather was ideal for photography. Optimism ran high on the ascent to that pretty home swaying so gently in the tree-top — and fell abruptly at the first glance into the nest! Some winged Assyrian had been down on the fold over-night and despoiled the nest, thereby bringing as much chagrin to a pair of spry youths as sorrow to a pair of brave birds! Earlier in the hunt for Shrike-Tits' nests I had endured with equanimity a fall from a tree-top, the breakage of a camera, and sundry other "incidentals," but this latest failure was the greatest trial of them all.

In good time, however, there came another opportunity for photographing my crested acquaintances, though not at the nest. In this case the home was altogether inaccessible, and had only been discovered through the agency of that powerful, Cuckoo-like pipe of the male bird. When the restless quivering of little wings above the rim of the nest could be perceived from the ground, I climbed some distance up and shook the slender tree-top. Instantly there was a startled chattering, and three baby Shrike-Tits fluttered out in different directions. Marking one, I followed it from place to place, shaking it gently from one sapling after another, until the poor wee thing came to the ground in sheer weariness. Then it was placed upon a low, horizontal bough, and tied down lightly with a piece of cloth (torn from the lining of my coat). And there the crested infant sat, "Charr-charring" wonderingly at the great world in general and its strange captor in particular, until its father flew down and sat alongside. That is how one of the accompanying photographs came to be taken.

Look, too, at the delightful solicitude betrayed by both parents when the family is yet unlearned in the ways of the great bush. They are just as proud of those shapely youngsters as the little Shrike-Tits — obviously more than content in the possession of that badge of notability, the crest, which with them is developed early — are of themselves. Let the slightest suggestion of danger intrude, and commotion reigns. With crests erect and tails spread wide, the parents flit anxiously about, uttering a harsh, chiding "Charr-charr," and now and again exploding into the piping whistle. Fired with the spirit of emulation, the young birds do their best to swell the medley, and chime in with a "Ta-ta-ta," "choo-choo-choo," the whole making quite an engaging chorus.

It would seem, too, that these juveniles are "supported" by their parents much longer than is the case with the average bird family. As late as mid-May of one year I was wandering about with a class of junior school-teachers when a strange note, which suggested either a young bird or a Robin in trouble, attracted attention. We searched for a while before a girl called: "See! It's got a yellow breast."

The next thing noticeable was the crest of a Shrike-Tit as the bird hung upside down, energetically cracking insects off the leaves. "Never heard a Shrike-Tit call like that before," said one; whereupon up flashed a handsome male bird, and fed the smaller bird as its wings fluttered in the tremulous style characteristic of the young "Yellow-hammers." It was seen again a few days later, looking after itself for the most part, but now and again trying to impose on hard-working adults by uttering quavering notes and quivering those eloquent wings.

At dusk of that same day a soft, Tit-like note led me

to inspect a thin, 16-feet high sapling. There, to my astonishment, I found *five* Shrike-Tits sleeping on the outermost fringes of the thin branchlets, and genially exposed to the rising moon. It was altogether a novel sight to see these active, vivacious birds perched so reposefully; and I went many times afterwards to the same locality in the hope of greeting the company again. But not until a full month later were those sleeping quarters occupied, and then only by one bird. Sitting bunched up, right out in an exposed position again, with crest for once at rest, what a close resemblance this solitary sleeper bore to a philosophic Yellow Robin!

In these cooler months the Shrike-Tits, like most other non-migratory birds, are moderately addicted to roving. Sometimes they drift into towns and cities. In 1914 a particular pair was in the habit of visiting the yard of a country fire-station, and the sympathetic keeper derived much entertainment from watching two Sparrows dancing attendance on them as they foraged in big gum-trees. When the strong-billed native birds ripped off bark, the Sparrows dashed in to share in the feast. But they got away again quickly, probably having an inkling of the fact that the Shrike-Tit is not sufficiently haughty to be imposed on with impunity.

I have seen a Shrike-Tit on its dignity many times, notably with a Yellow-tufted Honeyeater, each bird clinging to a slender branchlet and eyeing the other in a most quizzical fashion. In quite a different locality, on the brilliant morning of a June day, another meddlesome Yellow-tuft pursued a female Shrike-Tit, and a male bird whom she called up grew so excited over the affront that he used the piping whistle of nesting-time quite freely. How astounded he was when an interested human visitor joined in the monotone! Seeming to think a rival was in the field, he came down at once, with crest elevated, piping quite close to me.

A little later on there would be, in the natural course of events, disagreements among the Shrike-Tits themselves, just as there are among even the pure-voiced Bell-Birds — for even your most confirmed aristocrat must become "human" in the Springtime. On the morning of an August day a solitary female Shrike-Tit worked among bushes and called complacently "Kar-kar," while two fine male birds disagreed severely over her. It was not the usual straight-out contest. One bird seemed to be already in possession. He did not lead any assaults, but devoted himself strenuously to keeping off number two, who, with scant chivalry, kept driving at the lady. For half an hour I kept an unbiassed eye on this battle, and enjoyed it quite as much as, I suspect, did the cause of the trouble. At times both contestants grew vociferous, uttering shrill, chattering notes never before heard by at least one of the two spectators.

There are many other notes in the music-box of the Crested Shrike-Tit — among them a soothing little "croodle" of a tone which a human mother might adopt towards a restless infant; but over and above all is that round, piercing, distinctively Australian piping, a powerful, swelling note that does not seem to cost either bird any effort to emit. I have sometimes amused myself in imitating this call, and in noting the capriciously varied manner in which individual birds received the mimicry. A bird on a nest would usually peer inquiringly over the rim when it heard the signal, and nearly always a solitary female would respond thereto. A pair of nesting Shrike-Tits which dwelt away along the lower Murray,

THE ARISTOCRACY OF THE CREST

in South Australia, grew tremendously excited at hearing their family whistle in that lonely spot, and one of a trio to whom I whistled in the fastnesses of a mountain in Southern Queensland immediately threw back its head and answered the call.

But there were other occasions when representatives of the species were unimpressed, if not plainly bored, by such overtures. Mark the pride (and caprice) of the aristocrat!

Chapter VIII

DAYS AMONG THE ROBINS

To those who know, or have known, the Australian Robins "at home" — in their native haunts — the very name carries a strong suggestion of the free spaces of the land. There is a subtle *tang* in the word, something to make the heart beat faster at the impulse of memories of Winter days among the little scarlet gems of the open areas, and Springtime communings with the yellow-breasted nymphs of the woods. What a happy coincidence it is, too, that most of the scientific titles bestowed upon these lovable birds hold a pleasant significance! Moreover, if not all of them are musical, they have nothing of the "mouth-filling" appearance that causes some folk to wrinkle supercilious noses at certain Latin names.

Australia has, as set out by the Royal Australasian Ornithologists' Union, as many as 26 varieties of Robins, divided into eight genera. Comprising these are Red-breasts (4), Wood-Robins (4), Dusky-Robins (2), Shrike-Robins (12), Scrub-Robins (3), and one Fly-Robin. With the point as to whether any or all of these birds are true Robins according to the scientific standard I am not concerned; it is sufficient to know that they are of economic value to Australia, that many of them are considerably more beautiful than the Robins of other lands, and that some are as pleasantly fraternal as "the honest Robin that loves mankind" in the hard winter of the old world. By the same token, though, how useful it is for our Robins to be able to claim kinship with the estimable bird of Britain, giving them, as it does, a place in human affections that may not have been so readily attained on their merits!

For instance: A class of small school-children in Victoria was absorbed one day in a collection of museum specimens of birds, familiar and otherwise. The full interest of childhood was given every bird, but entusiasm was reserved for the Robin — the Red-breast of their nursery rhymes. And no repetition was needed of the invitation: "Tell me what you know about the Robin."

"Please, sir," said a small girl, "its nest is cold."

Here was a nice puzzle. Nesting Australian birds are more troubled by heat than cold. Then a recollection stirred to the occasion, and I recited gravely:

> Welcome, little Robin,
>   With the scarlet breast,
> In this winter weather
>   Cold must be your nest.

"Is that what you were thinking of?" An excited little girl said "Yes, sir!" and presently every sympathetic child was chanting in unison:

> Is the story true, Robin,
>   You were once so good

Mateship with Mother Robin.

"Talking birds" on the edge of a Queensland jungle.

Calling on the Tit-warbler,
Buderim Mountain, Queensland.

Calling on the Grey Thrush,
Gippsland, Victoria.

The Spotted Bower-birds' visitor, Central Queensland.

Marjorie and mother Dove.

"See!" At a Honeybirds' home, Central Victoria.

The dainty Sun-bird at home, North Queensland.

Well-hidden. White-plumed Honeybirds' nest in a dry bush.

Yellow-tufted Honeybirds' Nest. "The swaying type."

Yellow-tufted Honeybirds' Nest. "The supported type."

Fuscous Honeybirds' Nest.

White-plumed Honeybirds' Nest.
"A dainty, swinging cradle."

Happy Australians. Domesticated Cockatoo.

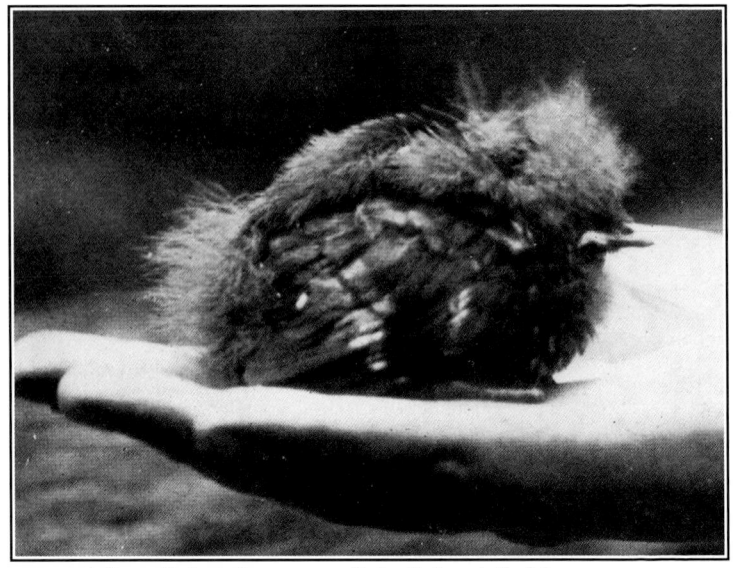

A "crest" that disappears. Baby Spine-tailed Logrunner.

The jungle home of the Whip-bird.

Edge of the jungle. Where Whip-birds and Shrike-tits meet.

A rare quartette. "The deserted homes were collected."

Nest of the Crested Bell-bird.
"Neatly built of bark."

Shrike-tits' nest.
"Swaying in the treetops."

The famous Whip-bird. "Jungle relative of the Shrike-tit."
(Photo. by R. T. Littlejohns.)

Shrike-tit and young. "Its father sat alongside."

Mother-love. Yellow Robin gazing down babe's mouth.

"How pretty!" Yellow Robin admiring its eggs.

"Look pleasant, please!" Yellow Robin "posing."

A rare study. Yellow Robin feeding mate.

A study in trustfulness.
Pale-yellow Robin on nest in jungle.

Watching Rose Robins. Mr. A. J. Campbell,
veteran Australian ornithologist, in a Victorian mountain gully.

Where Golden-breasted Whistlers abound. Queensland National Park.

Female Golden-breasted Whistler. "In a time of nesting."
(Photo. by S. A. Lawrence.)

Gilbert Whistlers' nest.
"I was able to photograph the nest and sunlit eggs."

Rufous Whistler at nest. "He came right back."

Rufous Whistler's mate. "The anxious little mother."
(Photos. by D. W. Gaukrodger.)

Photographing a Rufous Whistler's nest, Central Queensland.

Magpie-lark and family. "Face the world boldly."
(Photo. by W. G. and R. C. Harvey.)

The Magpies' Banquet. "Protected by their very prominence."
(Photo. by Sid. W. Jackson.)

Restless Flycatcher. "Sometimes confused

"The fighting spirit." Pied Butcher-bird.
(Photo. by W. G. and R. C. Harvey.)

Wagtail brooding. "A cup-shaped, cobwebbed home."

"A shady action." Wagtail protecting brood from sun.
(Photos. by D. W. Gaukrodger.)

White Watsonias on a Queensland Mountain. "The colour is that of the Butcher-bird's song."
(Photo. by H. Geissmann.)

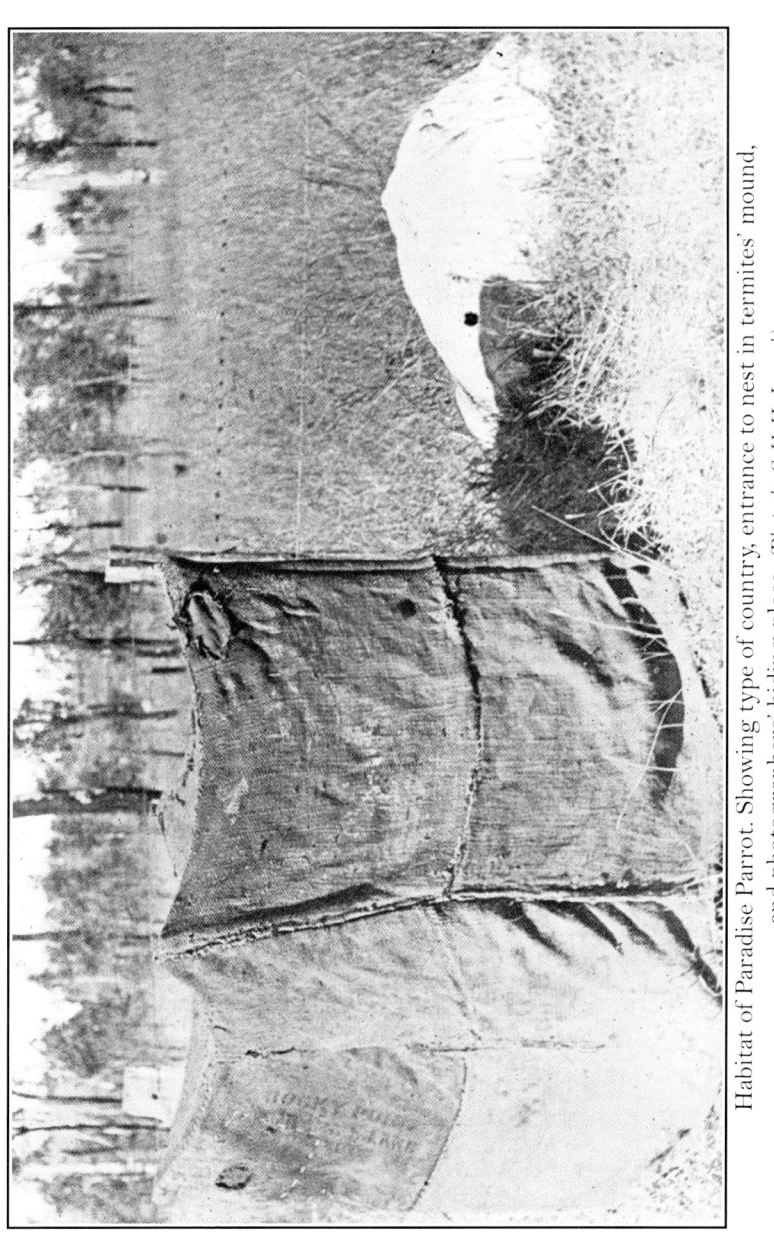

Habitat of Paradise Parrot. Showing type of country, entrance to nest in termites' mound, and photographers' hiding-place. *(Photo. by C. H. H. Jerrard.)*

Paradise Parrots; male at nesting-hollow, female above. "He would accompany her to the hollow." *(Photo. by C. H. H. Jerrard.)*

> To the little orphans
> Sleeping in the wood? . . .

Our most familiar Robin of the woods is not, however, one of the scarlet birds which grace the fields of winter, but a relative with a breast of jonquil yellow. Probably almost every school-child in southern New South Wales, Victoria, and South Australia is on more or less intimate terms with the Yellow-breasted Shrike-Robin — colloquially the "Yellow-Bob" or "Bark-Robin," scientifically *Eopsaltria Australis* (*Eos*, dawn; *psaltria*, a harpist), the Australian Psalmist of the Dawn. Among all the bird-studies that have appeared in magazines from time to time, probably no other species has proved to be so favored of our steadily growing band of bird photographers. The inference is obvious. Few among the smaller species of our feathered Australians are so readily located, and certainly no other is so "approachable," so well adapted to photography at the nest. It has what John Burroughs ascribed to a certain American bird, "civil and neighborly ways," and that practically the whole year through. The Yellow-Bob is no wanderer. It will remain faithful to one area amid all the process of the seasons, and frequently in spite of adversity.

When the mellow days of Autumn have given place to the severity of the southern Winter, when the migratory birds have all gone north and the nomads are wandering restlessly, the constancy of our small friend of the yellow breast and grey back makes his presence very gracious to the bush rambler. Go where you will, and you show yourself worthy the honor, the Yellow-Bob soon appears to give greeting. Sit down quietly in some bush recess, and it will not be many minutes ere one of these wood-nymphs has found you out. Then, with wings raised slightly, like inquiring eyebrows, he will inspect you from the vantage-point of the side of a tree — a favorite position for this bird — and, being assured, as like as not he will come and sit close by you for as long as you care to stay. There comes to mind the morning of a bright autumnal day, when I sat in a sun-streaked bush recess reading of "the ruddock with charitable bill" in Shakespeare's "Winter's Tale." And a philosophic little bird in yellow and grey, bunched upon a limb close by, betrayed the most profound interest in the whole recital.

Some naturalists have laid it down that a bird's silence invests it with an air of mystery. I do not find this to be the case with the Yellow-Robin. Certainly it is one of the quietest of our smaller birds, yet its pretty, trustful ways, combined with the eloquence expressed by its wings and tail, establish an understanding and sympathy between itself and the observer. Perhaps it is because of its voiceful performances at dawn and dusk that this bird remains comparatively silent during the full day. Apparently one of the unshakable rules of the whole genus is that its members must salute the morn and vesper the eve with a steady, melodiously solemn chanting. It was the hearing of this voice of Australia speaking "in the dim hour 'twixt dreams and dawn" that persuaded an unusually imaginative scientist to bestow upon the chief member of the genus a title immortalising the bird as the Australian Psalmist of the Dawn. And it is this melody that has helped to win for the Yellow-Bob a firm place in the affections of all lovers of Nature.

But not always is Robin silent in the fulness of daytime. A short time ago I heard one chuckling away to itself with all the melody, if not strength, of a soliloquising Butcher-Bird. In sooth, it is a rash thing to deny powers of song

DAYS AMONG THE ROBINS

to any of these little Australians on the superficial basis of our not having heard them lift up their voices.

Robin has a call, too — an intimate, thrilling "word" — more potent to move the heart of the bird-lover than any mere song, something as characteristic of the coming of Spring in Southern Australia as is the wandering voice of the Cuckoo in the Old World. You hear it first in the dying days of July, when the beautiful golden wattle is coming again to its all-assertive glory. It is the faint, yet resonant, nesting call, which, once heard, is associated ever afterwards with the haunting fragrance of wattle and the breaking of Winter's sway.

Soon, then, the pretty little bark homes begin to take shape. That these dainty nests are well known is not altogether the fault of their owners. For, in the majority of cases, they harmonise so neatly with the bark of the trees in which they are placed — as do also the pretty eggs with the green of the leaves hanging above — that it is easy for the casual rambler to pass them by unnoticed. Personally, I have discovered dozens of these homes, and almost always the cause has been familiarity with the interchange of signals between the birds as the unique feeding process is in progress.

Unlike most other species, the Shrike-Robins seldom take turn about on the nest. Their *menage* is based on an even happier arrangement. Mother does the brooding and father the foraging; and on this basis there is enacted a delightful domestic scene with each of the numerous meals — the more spectacular if there be babies in the nest. A beakful of luscious insects secured, the little lord clings in that funny Robin way of his to the side of a tree, possibly 15 or 20 yards from the nest. Thence he emits a faint note of inquiry. Immediately the quiet mother on the nest is all animation. Both wings quiver tremulously, and her pipe keeps rapid time with "Me-me-me-me, yes please," and the notes are cut off sharply as the Psalmist flits up swiftly and silently, places the offering in her bill with astonishing rapidity, and is away again as quickly as he came. The wings in the nest that were so "troubled and thrilled with ecstasy" are folded quietly for a brief space as the mother bird "prepares" the morsel in her beak; then she stands, hops back to the edge of the nest, and feeds the young.

The camera can record much, but it must ever fail to re-create the moving, loving expressions of the mother Robin as she spreads her wings over the babies, snuggling and snuggling until they are perfectly comfortable. "Bairnies, cuddle doon," the pretty actions plainly say, and if a bairn is unduly restless the parent will stand erect on the edge of the nest, look fondly down at the little one, and deliberately soothe it into quietude. Over many years I can see still the devoted mother of a Robin family to which I first gave close attention, photographic and otherwise.

I made their acquaintance towards the end of September, the mother then being in proud possession of a nest neatly and protectively built into a fork created by a branch shooting out from a trunk of a box-tree, and having the usual nest-carpet of small dry leaves. Early in October one of the two handsome eggs gave forth life, and the useless one was cast out. Possibly it was because they had only one babe to attend that the parents grew more solicitous for its welfare than is customary even with Robins. From the moment that little black ball of quivering flesh stirred beneath her the mother became a fanatic, and displayed little fear of the inquisitive eye of a stand-camera placed below the nest. Too much movement would cause her to flutter away, but very soon she would be back, guarding the treasure again.

In the course of time the camera secured an example of the bird crouching, a clever device no other species can exercise so well. Sitting on the solitary young one, the mother would be moving her head in all directions in anticipation of possible danger, when the crackle of a footstep on dry leaves would come faintly to her. Apparently it was quite obvious to the little creature that her back was protectively colored grey, while the yellow of the breast was noticeable, for instantly she would sink flat down in the nest until only the tip of the bill and a bright eye were visible above the rim. It was a matter for wonder that the bird could efface herself so completely. What was more curious still, the knowing little thing always distinguished between human and animal steps; at the soft pad of a dog she would start up and peep inquiringly over the edge of her bark dwelling.

It was worth while, too, walking close to the home and looking in another direction. Then the mother would hop swiftly off, drop almost flush with the ground, and fly softly into the scrub. But never did she go far away, and, directly the little one gave a call of alarm, the brave parent would go almost frantic with anxiety, and be back putting up a splendidly natural imitation of a young or disabled bird. The Chats, the Ground-Thrushes, and one or two of the Honeyeaters are artists at this ruse, but none of them can spread the feathers and assume the fluffy, babyish appearance of the mother Robin. The effect created is quite pitiful, and more so when the distraught little bird stretches her wings to drag along the ground, and, anon, raises them till the primary feathers touch above the back. As soon, though, as the danger passes, the small actor is herself again, and, on the instant, is *cuddling* her little ones in that typically tender manner of the Robins.

Probably these sprites remain faithful to the one partner while life lasts. I have seen three birds feeding a single brood of potential Psalmists, but it seemed to me that number three was a lonely member of an earlier brood, practising on his baby brethren.

It was a pair of Yellow Robins which nested in the same district as those particularly devoted parents that provided the most remarkable instance of pertinacity I have experienced in Birdland. Early in August of 1912 they built first in a thin sapling. The time was too early for such a precarious position, and the high winds blew the frail structure sidelong. With a piece of pliable sapling I tied the home up and saved the precious eggs, whereat the mother came back at once, and sat serenely, until the next strong breeze blew the insecure structure to pieces. But the material was not wasted; the birds gathered it up and built, more wisely, in the fork of a sturdy sapling some twenty yards away. Here they were safe from the winds, but not from other dangers, and presently there were *two* of the pretty pink eggs of the Bronze Cuckoo to keep the green ones company. Some birds throw the Cuckoos' eggs out, but this pair accepted them in good faith, and soon a young Cuckoo was hatched out. The interloper would probably have thrown the Robins' eggs or young ones out very soon, but for another development: a marauding bird took the lot.

Still the plucky Robins stuck to their chosen locality; they built again, low to the ground, at a point about ten yards from each of the previous sites. But once more they were doomed to disappointment, either a reptile or a small boy — they are equal in point of pestiferousness in this regard — confiscating the third set of eggs.

It was November by this time, and the Psalmist and his indomitable wife must have been growing dubious about the desirableness of their chosen locality. However, they emulated the Bruce's spider — tried again, and succeeded. The fourth nest was built opposite the third, and the whole quartette was within a radius of forty yards. By January a proud pair led forth a lusty brood, and their confidence in the locality was sufficiently restored to persuade them to nest there again during the ensuing Spring.

Our familiar little friend is the only one of its kind in the south-eastern portion of the continent. At some indefinite point in New South Wales the species blends into a prettier relative, in whose case the yellow of the breast is continued on the lower portion of the back. Then, in the coastal district of central Queensland, the brightness of the rump fades, and a replica of the southern bird, albeit a trifle smaller, appears, to be replaced in turn by another brightly marked species in the jungles of the north.

Why this curious undulation in color?' With such markings as the orange wing-patch of the Tree-Creepers, the yellow rump of the Tit-Warbler, and similar features of other birds of the open areas, the Wallace theory of "warning colors" may well apply — and it would seem that yellow is a distinctive color to the eye of a bird. But the consideration must be carried further in the remarkable case of the Robins, where the bright patch is possessed only by those birds which dwell in the thick jungles and dimly lit gullies. As in the other cases quoted, this feature is noted most distinctly when the bird is flying; but the Robin's quaint habit of clinging to the side of a tree, with its back to the visitor and wings poised inquiringly, makes the golden rump to be noticeable even when the bird is at rest.

This fact became vividly apparent on an evening when I watched a pair of Shrike-Robins in a tea-tree gully near Brisbane. The flitting forms of the birds could hardly be seen in the half-darkness, but when they clung to trees the golden rump could be discerned with a strange clearness. It seemed, indeed, to impart a certain glow, not unlike the phosphorescent "light" which skirts the mouths of young birds born in dark places. And so I assume that patch of gold to be not so much a danger signal as one of Nature's beacons, an attribute unnecessary to birds which live in the sun.

The observations that have been made with regard to the friendly Yellow Robin of the South apply in the main also to its prettier relative of the North. The nests and general housekeeping arrangements are alike — the male feeds the female on the nest — and there is the same questioning flick of the tail, tilt of the wings, and sharp "Clip, clip" of the wings when the bird is making a flying inspection of a newcomer. Here again, too, you get those neighborly, sociable ways, even though the habitat of the bird makes it rather more of a recluse than the better-known Robin of the South.

The first occasion on which I met the Yellow-tailed Robin is remembered well because of a clever flanking movement on the part of the bird. Leaning against a large tea-tree, I lost sight of my new-made acquaintance as it described a semi-circle, and turned soon to find a pair of round eyes, bright with inquiry, making a close examination from a few yards to the rear!

Mark, too, the almost startling experience of a young settler residing on a mountain in south-eastern

Queensland. In all soberness, he relates that, as he stood panting one day after felling a tree, a Yellow Robin flew up, clung to his lip, and pecked at the white teeth. The story is too good to be doubted, especially when one remembers the penchant which young birds have for pecking at teeth when being fed from a human mouth.

Taking the other varieties of Robin by and large, the first point to strike the student is the uneven and curious nature of their distribution. Why, for instance, should Tasmania, which has no Shrike-Robins, also be minus Scrub-Robins, Black and White Robins, Red-capped Robins, and Rose-breasted Wood-Robins? May we take it that the island State was separated from the mainland before these particular branches of the family were evolved? The puzzle is the more profound in respect of the two birds last-mentioned; for the pretty Red-cap is a close relative of the Scarlet- and Flame-breasted Robins, both of which are commensurately more plentiful in Tasmania than on the mainland, and the rose-breasted sprite of the woods is allied to the Pink-breasted Wood-Robin, which is also fairly plentiful in portions of the island State.

Moreover, apart from the question of definite distribution, what a lot there is to be learned concerning the seasonal movements and general wanderings of certain of these birds with the roseate breasts. The Red-capped and Scarlet-breasted Robins may be passed over lightly. There is no particular mystery about their comings and goings. They are two of the three species of red-breasts which create such beautiful patches of color in the green fields of Winter. Perhaps the Red-cap is the more widely distributed. Certainly it leaves Tasmania out of its orbit, but it is found in many of the drier out-back portions of the mainland, where other Robins seldom, if ever, venture.

And, in Springtime, the nests of both of these species have been freely found in various parts of the eastern States.

Not so in the case of the Flame-breast. Its nest has rarely been found outside of Tasmania. Indeed, the showy bird itself has not often been seen in Australia proper other than in the cooler months. Towards the end of April of each year the females of the species — unobtrusive brownish birds, with a touch of white in the wings — arrive unostentatiously in the open areas, and, within a week or so, their gaily dressed lords follow. No one ever sees them actually arrive. No one ever sees them recommence their journey. Nevertheless, flocks of Flame-breasts have been observed, several times, en route from the mainland to Tasmania, and so the belief has become established that, on some obscure prompting, these little creatures (whose wings are not made for long distances) journey annually backwards and forwards over Bass Strait, and do not, as John Gould had it, "retire to the forest to breed."

To some extent, however, this theory is discounted by the fact that a fair number of Flame-breasts are to be found in Tasmania all the year round. At the present stage, then, the indications are that this one species of Robin does actually undertake this annual "sea trip," but that, in each Springtime, a certain number of Flame-breasts elect to nest on the mainland, and that a larger number still spend their whole time in Tasmania.

The housekeeping of all the little *Petroicas* is on much the same lines. They build dainty nests of bark and fibres, half the size of the homes of the Shrike-Robins. The usual site is a tree-fork or the cleft of a stump, but frequently the Flame- and Scarlet-breasts choose the upturned roots of fallen trees. The length of time required for the males to assume their gorgeous livery is probably two or three years. But they do not wait for that before mating; it is not uncommon to see the birds nesting in adolescent plumage, when the male is indistinguishable in color from the female.

The Pink and Rose-breasted Wood-Robins (*Erythrodryas*) were formerly grouped with the Red Robins of the fields. They are, however, quite unlike these in many respects. Their habitat is the semi-humid gullies of the eastern and southern coast, and, whereas their relatives wage war against the ground insects, these small wood-nymphs take most of their food on the wing. Here, again, the problem of distribution and movements is a real one. These kin-spirits, who are so much alike that they were confused for many years, breed together in the gullies of Gippsland. After that they part company. The larger Pink-breast is found nowhere but in Victoria, South Australia, and Tasmania, while the tiny Rose-breast spreads northward from Victoria, through New South Wales, to Queensland.

About the middle of April of each year I go to a certain tea-tree gully — in the neighborhood of Brisbane — to welcome back the Rose Robins. If they are not there one day they will be there the next, announcing their presence with a faint, tremulous "He-e-e-re." Watch closely, then, and you will see a faint flutter of wings in the tree-tops, and presently a rosy-hued male or grey-garbed female will come into view, twinkling gauzy wings in a remarkable fashion typical of the species. No other bird known to me has this pretty habit developed to such a pronounced extent. The quivering of the wings affected by the Shrike-Robins and Shrike-Tits is a thing apart, an ecstatic motion born of nesting-time; moreover, the rose-breasted birds of Autumn do not so much quiver their wings as impart to them a *winnowing* motion, now stretching them above the back, now drooping them about the feet, while filtering light gives to the extended feathers a delicately translucent quality. And when the bird takes flight from tree to tree it resembles nothing so much as a long-tailed butterfly.

I hope I have not developed into the officious showman in respect of my friends the bush birds, but there is always delight in taking an appreciative visitor to make the acquaintance of this particular Robin, whose beautiful presence is one of the chief features of the tea-tree glades in Winter — that is, for those whose eyes and ears are sufficiently skilled to catch the faint, insect-like chirp and slender little form. There was a Sabbath afternoon in May when two grave and reverend signiors, one a Scottish divine and the other a University lecturer, found themselves wandering on and on through a tea-tree glade, and enjoying, quite as much as an ordinary bird-lover, the sermon preached by the Rose Robins on the wisdom of being chiefly merry and bright.

If the Rose Robin is essentially cheerful, that is not to say that the bird is a model of amiability. On at least two occasions I have seen a male and female of the species sitting lengthwise on a horizontal bough, and "Churr-churring" at each other in most animated fashion, the bright-breasted bird apparently being the moving spirit in each case. On another day, as I walked along a road skirting a tea-tree gully, attention was claimed by a strange bird-note, a sharp "Tick-tick," as of a twig snapping; and presently there flashed into view a beautiful male Rose Robin, fleeing ignominiously from an excited little female of the species!

Again, on August 31, 1918 — the date is notable as the

latest record for *Erythrodryas rosea* amid the tea-tree — I was watching a lordly little Scarlet Honeyeater coming down to a pool, when he was forced to flee from the onslaught of a pugnacious Robin of the opposite sex, who then turned her unladylike attentions to an inoffensive Tree-Tit. This particular Robin was the most consummate of all wing-artists, extending the pinions wide rather than drooping them low about the feet.

Whether the females of the Rose Robins (as in the case of the Flame-breasts) are the first to lead the way on the annual pilgrimages cannot yet be said with any certainty. I was inclined to return an affirmative answer to the question until the year 1919, when I saw my first Robin as early as April 13, a bright-breasted male bird, which sat quietly in a leafy sapling, as though tired out after a long journey.

With the coming of September the little drooping wings are no longer seen in these Queensland gullies. The birds have moved off in the warming nights of late August, and are then working their way down the coast to southern New South Wales and the recesses of Gippsland, or up into the coolness of the coastal ranges. In the absence of ornithological evidence on the point — for it is strange how little has been written of this fascinating bird — the non-discovery of Rose Robins' nests in Queensland seemed to indicate a distinct southern migration with each Springtime, just as it was formerly believed that the Flame-breasts of the mainland crossed to Tasmania with the passing of Winter.

The finding of several pairs of Rose Robins dwelling on the shaggy heights of the Macpherson Range and Bunya Mountains (South Queensland), during the latter part of the year, has caused a modification of this view. Interstate migration may be practised by the Robins, but certainly it is neither complete nor general. How strange it is, then, that Queensland yet lacks a record of the discovery of the dainty nest of little *Erythrodryas* of the rose breast! Conversely, too, is it not curious that the species is not remarked on during the Winter of Victoria?

For here in Southern Queensland the birds are not at all rare, either amid the tea-tree or elsewhere, during the cooler months. Indeed, I hear the queer little call at this time in almost all classes of country about Brisbane, and for at least 100 miles farther north, sometimes nearly a score of miles inland, and sometimes right on the margin of the sea.

Chapter IX

FINE FEATHERS AND FINE BIRDS

Whatever justification there is in a world of men (and women) for the popular belief that fine feathers make fine birds, there is more of sound than wisdom to substantiate the catch-phrase in relation to the birds themselves. In any case, you must first be sure of the relative value of the adjective.

There are those fortunate souls who rejoice in the view that all birds have their fine points, the only distinction to be drawn being a matter of degree. They will direct you to the fact that the absence of bright feathers matters not at all to birds of consistently plain dress, in which class are included original merry-makers and jesters and the most distinguished of the world's songsters — fine birds, to be sure!

Nevertheless, it is not safe to dogmatise in another direction, as poor Lindsay Gordon did in striking that false, still-echoing note regarding the "songless bright birds" of the land of his adoption. That many of Australia's song-birds are modestly garbed is true enough, but there are several others gifted with plumage no less rich and challenging than their melodious calls, whistles, chatter, or songs. In this regard I think particularly of certain members of the genus *Pachycephala*, generally known as "Whistlers."

Australia would be a good deal poorer ornithologically without many of the seventeen or eighteen species which constitute the Commonwealth's representation of this, a large group "peculiar to Australia and adjacent islands to the northward." I use the word *many* for the reason that at least half of our full number of Whistlers are confined to the far north of the continent, and are unknown save to the pioneering ornithologist, who, gun in hand, will brave the dangers of the mangrove swamps and other waste places on the chance of discovering a "new" bird.

But there are others of the family which take rank among the common birds of Australia, and which would be even better known if they had better names. As with several other Australian birds, their vernacular title is distressingly indefinite in a country "where the very air is ringing" with the melodious whistling of full many birds. This much may be conceded, however: "Whistler" is infinitely to be preferred to the old term of "Thickhead," which some ungracious scientist, not satiated with the generic title (*Pachus*, thick; *kephale*, the head), early bestowed upon the birds — because, forsooth, dissection showed them to be comparatively broader across the cranium than most other families!

Theories on the point of sexual selection and protective coloration become somewhat involved in considering Australia's Whistlers. Practically all of them are possessed of sweet voices, but they show broad distinctions in coloration, from sober grey to bright yellow. Half a dozen species, in fact, are favored with plumage which ranks its owners among the beauty-birds of the world. Were these bright-colored Whistlers originally of modest dress, gaining their sunset yellows and reds in the competitive struggle of the ages? There are connecting links of color and voice to support this idle presumption; but then, why should certain members of the group retain their sober garb while kinspirits were growing in beauty?

Almost it would seem that the grey birds are sensible of an invidious distinction, for theirs are the plaintive voices among a party of notable melodists. John Gould described the notes of *P. simplex*, a little-known brown Whistler of the far North, as "peculiarly soft and mournful," and, listening again in fancy to the softly Celtic tones of a grey bird in the South (*P. gilberti*), I find both adjectives quite fitting. In each case known to me the voice seems to accord strangely with the garb of the bird. Surpassing sweetness is an ever-present quality, but, whereas the calls of the quietly dressed birds are sweetly subdued, those of the golden-breasts are rich and challenging, while the sparkling lilt of the warm-brown-breasted birds achieves the dignity of a song.

For all their gay clothes and disposition, however, the male Golden- and Rufous-breasted Whistlers, equally with the paternal birds of the less noticeable species, take their turns at brooding the eggs and caring for the babies. Here, again, the puzzle of coloration is deepened. If the female bird is quietly colored for protective purposes, why does the male assume charge of the home? Is the grey or dull green coloring of the back — so general a feature with bright-breasted birds — sufficiently neutral to ensure him obscurity, or, remembering that these birds are known to breed while yet in adolescent plumage, was the usefulness developed before motley became the only wear?

I have no personal knowledge of the length of time necessary for the attainment of the male Whistlers' regal colors, but am quite prepared to believe that at least two years must pass ere the beautiful gold and black emerges from the confusing changes of juvenile plumage of the birds with the sun-flower breasts.

These, undoubtedly, are the beauties of the Australian portion of the family. Picture a plump little bird colored grey on the back, black on the head, and white on the throat, with a frontal expanse of rich golden-yellow separated from the throat by a black band, and extending in a golden half-circlet around the back of the neck; imagine these colors blending harmoniously with one another and with the chaste green of a young eucalypt, and you have an idea of how striking a living miniature any one of the male Golden-breasted Whistlers can present when in full plumage. Ornithologists divide these particular birds into at least five species, namely, the Golden-breasted Whistler (*P. gutturalis*), common to Queensland, New South Wales, and Victoria; the Black-tailed Whistler (*P. melanura*) of the North; the Grey-tailed Whistler (*P. glaucura*), of Tasmania; the Southern Whistler (*P. occidentalis*), of South Australia; and the Western Whistler (*P.occidentalis*), of West Australia. And no one but an ornithologist would see any difference between the lot! The assertive gold, white, and black is present in each species, variation showing almost solely in the coloration of the tails.

All this display of rich plumage renders the title of "Golden-breasted" Whistler for any one of the gorgeous bunch doubly indefinite; but, in the absence of a more suitable name, the most familiar of them is entitled to chief distinction for having been, so to speak, first in the field. ("The *Pachycephala gutturalis*," wrote the veteran

John Gould, "may be regarded as the type of this genus.") The only other title worth noting as applied to the Golden-breasted Whistler is that of "Thunder-bird," early bestowed by the colonists of New South Wales, who noted the beauty bird's pronounced penchant for breaking into a challenging reply to a growl of thunder or the report of a gun. But other Whistlers will do the same thing; it is a curious little trait of the genus, practised by the soberly clad birds equally with their more haughty relatives.

Be the mood of the Golden-breast what you will, however, the notes are never aught but sweet. What the strain lacks is continuity, and, in a lesser degree, variety. Still, occasionally an individual bird may be heard in quite a genuine song — like the opening notes of a gavotte, said a dweller on a Queensland mountain, as we listened appreciatively to a bird of unusual powers. These jungle areas are much favored of the Golden-breasted Whistler, but I have nowhere found the species more plentiful than in the far east of Gippsland, bordering on New South Wales. That pretty quarter rang with proud, unvaried calls of the spanking male birds in a time of nesting, while the more sober-minded females tended eggs or young in the shade of the tea-tree.

You may see this little lady, perhaps not to better advantage, but certainly on more intimate terms, in the Autumn. The sexes appear to separate at the end of Summer. Time after time I have watched solitary males and females respectively, but only on very few occasions have I seen an apparently mated couple between the beginning of April and end of August. And yet no individual bird is ever lonely; if so, at least the sentiment is well hidden. Each one spends the quiet months profitably to others beside itself in working among the leaf-insects of the eucalypts, and a happily reflective whistle between whiles announces that all's well in the immediate world.

An indication of the bird's presence is given by the constant "Crack, crack" in the trees frequented; only, it is well to be sure that the busy Shrike-Tit, which much resembles the male Golden-breast, is not the responsible party. By reason of his coloration, the male Whistler is easier to locate than the female, but I have found him to be more of a wanderer. During the cool months of the past few years I frequently met the handsome bird, but always by chance; whereas there were at least four gullies in which I could depend on finding a female of the species — just one to each area.

That is to say, the bird was constant to the one spot the Autumn and Winter through, but the *finding* was a different matter. Creatures of unusually curious impulses, the little grey ladies sometimes remained quiet for hours at a time, and on other occasions waxed melodious. Midway in the month of April one of these birds emitted its rich, spasmodic call — "Whee, wee-wee!" — and then came down and "charr-charred" brazenly at me, so much in the tone of a chiding Yellow Robin that one of these birds excitedly flew up to investigate. After that it became apparent to me that the grey-garbed bird with the touch of red in the wings was an autumnal melodist of no mean order.

It is not straining at a fancy to say that the bar most frequently used by the solitary female wanderer suggested the words "Be quick, quick, oh-please-do-be-quick." Silence for a variable time, and then the strain changed to: "Swee-ee-t, swee-ee-t; oh it's pretty, it is pretty — *pretty.*" Almost every bar was preceded by the curious, "half-indrawn" whistle characteristic of the genus — a note

FINE FEATHERS AND FINE BIRDS

suggesting that the performer was gathering breath for the effort. Then, more rarely, there was a remarkably rich bar: "Bobby-link, bobby-link, bobby-link, bobby-link."

The bird heard to best advantage on this rollicking note came down almost within arms' reach to inspect me; at other times it was impossible to persuade Miss Caprice from the tree-tops. She would simply screw her head on one side in that quaint fashion which is another trait of the Whistlers, and listen to the whistle of invitation with a detached air, as who should say: "Lord, what fools these mortals be!"

This is a typical note taken on one of those Autumn days: —

> Heard a slight, sweet note in the gully, and presently found another of the remarkable female Whistlers. These birds are evidently all of the one mind, in that each is always alone, always in a timbered gully, talkative only on odd occasions, and alternatively curious and shy. In two hours the bird emitted but one rich note, quite different from all others I have heard.

Two weeks later, on the morning of a clear June day, I was intently watching the curious spectacle of an arboreal Shrike-Tit working on the ground, when the shrill "Peeeee!" note caused me to look up quickly. There, in all his glory, was a brilliant male Golden-breast, darting from tree to tree around the female Whistler of the locality. It was as though the fire of Spring was already in the air. "Seeeee!" he called, in a prolonged, ecstatic note, as he flaunted his gay colors for admiration; then, as no response was forthcoming, "Be quick, quick, *quick*!" And still the magnificent indifference was maintained; the small grey bird continued to feed quietly, her whole attitude suggesting the amiable scorn of a Beatrice: "I wonder that you will still be talking, Signior; nobody marks you!"

For several minutes this little comedy was kept up, the beauty-bird dashing all around the object of his affection, and uttering the shrill note and the melodious "Be-quick." Betimes he went too close, whereupon a sharp, unladylike peck helped to cool his ardor. Antagonism added to indifference was too much for the proud wooer; he discontinued his advances shortly afterwards, and, when the saucy grey bird flew off, he simply sat preening his golden feathers, and did not attempt to follow.

Possibly the lady Disdain returned to look for the gay cavalier when a realisation of her loss dawned upon her, but, again — I say this quite seriously — the incident may have been only a passing flirtation. If the Whistlers are at all consistent in the apparent separation of the sexes during the cool months, one would hardly expect them to pair up as early as June. In Queensland I have seen two beauteous male birds calling each other strange names over the one untroubled female, but that was near the end of July.

Leaving the birds of the golden breast, chief among other members of the clan, in point of wide distribution as well as in brightness of voice and plumage, is the familiar, sprightly, and altogether lovable bird known as the Rufous-breasted Whistler. As in the case of its regent-breasted relatives, the name has purely to do with the male bird, the grey-clad female receiving it only as a matter of form. Nor are there other Rufous-breasts to dispute the title, the nearest relative being *P. falcata*, a Whistler of the extreme north, which possesses the same color scheme, but of a much duller quality. For the rest, the Rufous-breast is the same in every State in Australia, but, strangely enough, Tasmania is excluded from its itinerary. Albeit much more definite in its movements than the golden birds, the Rufous-breasted Whistler is a nomad rather than a migrant. You may chance to see either the pretty male bird, clad in reddish-brown vest, grey coat, and white collar with black band, or the quietly garbed female bird in almost any part of Australia at any time of the year. Pre-eminently, however, the bird is one of the elves of Spring.

In Southern Australia Spring would not be complete without the Rufous-breasted Whistler. Its clear, ringing warble is one of the most joyous lilts in the bush and country towns from early August to the end of the year. All libels to the contrary, the female bird is scarcely less melodious than her eloquent partner. It is the rufous-breasted bird, of course, that chiefly furnishes the well-spring of music, but I suspect this to be due largely to the fact that he has, in the words of Prospero, "more time for vainer hours" than the expectant mother-bird. Into the paean of Spring the pretty visitor throws his whole spirit.

There is one bar in particular, a chattering ripple suggestive of the pattering of elfin feet, that an individual bird will sustain for as long as half a minute without pausing for breath, what time the throat pulses and the little body vibrates with the melody. It is, too, a notable fact — and here is a feat which no human singer could safely attempt — that the spontaneous music of the birds is not hampered by a full mouth; when photographing young Whistlers I have seen the parent birds emitting passionate protests from bills that were almost crammed with orchard flies.

It is a predeliction for orchards that has earned for the valuable bird the name of "Gardener" in some parts. "Joey-joey" was another home-made name which I liked better, not merely because it was announced by the bird itself in the series of running, tumbling notes that follow the initial whip-crack, but by reason also of the affectionate way in which the bush-woman who used the title tossed back his melody to the joyous bird in the early Spring. From the same notes, I surmise, came another fraternal title of "Joy-Bird" — with which there is nothing out-of-place but the fact that nearly every bird in the land does its little best to deserve it.

Probably, the most widely used title for the Rufous-breasted Whistler is that of "Bush Canary," though here again your average observer creates a genial confusion by applying the same title to various bush melodists, notably the dear little white-throated Warbler. No other bird, however, could well have been the subject of these lines by Arthur Bayldon:

> A Bush Canary — hark, oh hark!
>   His raptures fill the vale,
> Kin-spirit of the frenzied Lark
>   And sobbing Nightingale.

To a more recent poet, the reverend author of those good bush verses, "Around the Boree Log," the Whistler is the Wiree (wiry), a quaint little title drawn from the "lilting lay" of the brilliant bird:

> "*Wir-ee, Wir-ee, Itchong, Itchong,*"
> Then rippled through its liquid song.

Henry Clarence Kendall heard a similar song over fifty years ago. His reference to "E-chongs" marked the entrance of the Rufous Whistler into its country's verse.

It is, I think, fair to assume that the nomadic "Joy-Birds" are constant to the one favorable locality for breeding-purposes. Year after year a pair came (returned?) to a favorite bush orchard in the bright days of early September. No one ever saw them arrive. One day there would be no hint of their presence, and the next day the old orchard was vocal with melody. Given a week or so for song and play, the minstrels would commence housekeeping. The fragile, fibrous nest involves very little constructive labor, and most of this is done by the female. Her gay consort, however, is not at all backward in tending the queer, olive-colored eggs; nor is he at all lacking in fidelity to the babies.

In nine cases out of ten the mother-love of a female bird renders her distinctly the braver, not to say the more assertive, when the young are apparently menaced. (This, of course, explains why it is the mother-bird that appears most often in photographs of birds' nests.) It was not so with the family of Rufous-breasted Whistlers which afforded my first pictures of the species. In the heat of November the parents were briskly attending a pair of hungry offspring in an orchard home when the ubiquitous photographer found them out. Not even the strange-looking eye of the camera could daunt that dutiful male bird. He came right back to those ravenous babies while the focus was being obtained, thereby setting an example which the anxious little mother was not particularly slow to follow.

With the coming of the new year the fire of Spring has died out of the revelling Whistler, though, to be sure, he does not altogether exhaust the throbbing melody of old. Occasionally you may hear quite a rich canticle in the peace of Autumn, but "never a one so gay" as that joyous greeting to the Spring.

We leave now the birds of striking plumage for a relative whose dress and disposition are infinitely more modest. It does not seek the society of man in the fraternal way of the Rufous Whistler; nor has it the general distribution of the birds of the golden breasts. John Gilbert, the able English naturalist who was killed by North Queensland blacks (when with the Leichhardt expedition of 1845), met with this sober little bird in the dry Mallee areas of Victoria and South Australia, and Gould paid his tireless assistant the compliment of naming the species after him — *P. gilberti*, the Gilbert Whistler. This use of proper (human) names in bird-nomenclature was not, thanks be, a failing with Gould, but Gilbert was worthy of the honor.

In color the Gilbert Whistler is closer to the reds than the golds among its kind, the generally greyish tint of the plumage of both sexes being relieved by reddish-buff on the throat of the male bird. But in its housekeeping it more resembles the Yellow-breasts; the Thrush-like nest is much more substantial than the rude home of the Rufous species, and the pretty eggs, with their pattern-rings of brown and blue spots on a white background, approximate to those of the Golden-breast. If there is any application to birds in the declaration that "by their fruits ye shall know them," the scientists are wrong in placing Gilbert's species next in the group to our familiar friend, *P. rufiventris*.

The late Mr. A. J. North, of the Australian Museum (Sydney), described the Gilbert Whistler as "the rarest species of the genus inhabiting the southern portion of

FINE FEATHERS AND FINE BIRDS

the continent." Its range extends from the north-west of Victoria across to West Australia, branching southwards to the centre of the former State.

It may be that this southern extension is a semi-migratory movement of the Springtime; for that was the only period in which I met the sweet-voiced bird outside the barren portions of the Mallee country. My acquaintance with the species "at home" dates back to 2nd October, 1912. Cycling slowly along an old bush road in central Victoria on that day, I glimpsed a large bright eye of a bird peering over the rim of a nest tucked into a bush-covered stump about three feet in height. It suggested the Grey Thrush, but closer inspection showed the bird's bill to be comparatively small. When flushed from the nest the stranger's identity became apparent.

The home was finely built, chiefly of grass, most compactly and neatly matted into a round wall. Presently the male bird appeared, uttering a low, plaintive whistle of alarm. Then both birds, anxious for the safety of the two pretty eggs, kept flitting around from bush to bush, each emitting an exceedingly sweet call, sounding as "Wee — , woo — ," the second syllable dropping with a pretty mournfulness. Certain other notes resembled some used by the Rufous Whistler, the whip-like crack being even stronger. It was preceded and followed, too, by a soft, sweet note that suggested an echo of the crack coming from far away.

On my next visit (7th October), the male bird was in charge of the nest, and all was well. Two days later the female graced the home, brave in the pride of motherhood; a wriggling, broad-headed little bird had emerged from one shell, and the other was suspiciously dark. During the next few days the solitary chick — the second egg proved infertile — thrived; but on 19th October there was an empty nest and wailing parents. Six days later the nest had been removed — taken with a cleanness indicating that the birds were responsible.

After that I heard but little of the Celtic-voiced birds until September of the following year. On the morning of a dull, close day the ventriloquial notes echoed softly about the same locality, and, less than a fortnight later, I found the nest. It was built neatly into the top of a collection of sticks that previously had done duty as the nest of a pair of Babblers. The female Whistler fluttered off at the sound of a foot-fall, and did her brave best to draw attention away from the precious eggs. Hopping stiffly over the ground, she "fluffed" her feathers tragically, not in the wild, distracted manner of ground-nesting birds, but exactly after the style which the home-loving Yellow Robin often adopts to lure away intruders. Though like in cause and effect, this pretty performance is quite distinct from the "broken-wing" ruse used by the White-fronted Bush-Chat, Yellow-tufted Honeyeater, and one or two other species. Two days later the eggs had disappeared, presumably having been stolen by marauding boys, and the site was deserted.

Evidently the Whistlers do not take long to build a nest, for within eight days the birds were at home on a bushy stump less than 100 yards from the site of the Babblers' nest. The new nursery contained two beautiful eggs — soft-cream, spotted with blue and sienna. I was able to photograph the nest and sunlit eggs, but the owners could not be persuaded to return while the camera was in position. Timid little creatures at any time, their faith in human nature had evidently been badly shaken by rough experiences.

FINE FEATHERS AND FINE BIRDS

It was worth while, however, to spend many hours in the vicinity, if only to hear the airy wood-music of the birds. The call most frequently used was a ventriloquial "Chu — p, chu — p," which seems to roll softly off the chest and swell powerfully as it leaves the beak. So faintly does this chant commence that you may imagine it to be coming from one hundred yards away, until, as the sound gathers rapidly in volume, the author is located close at hand. Here again, as in solicitude for the home, the Whistler shows its kinship with the Shrike-Robin; for this call is distinctly reminiscent of the vesper hymn of the familiar Psalmist of the Dawn.

A third egg was duly added in the second home, and housekeeping proceeded smoothly for a period that was, alas, only too short. On 12th October one of the trio of eggs was gone, the other two were cold, the horse-hair lining of the nest was ruffled, and the birds were calling in melodious pain two hundred yards away. They came no more to that nest, but clung with resolute loyalty to the chosen locality. It was all of no avail. On the last day of that month I found an empty nest, without any signs of young birds having been in it. There seemed an additional touch of plaintiveness in the Whistlers' melody then; they had been thwarted for the third time! For another fortnight the sweet voices sounded rarely in the vicinity, and then were heard no more.

It is part of the fascination attached to "mateship" with the birds that the caprice of the little creatures continually creates pleasant surprises. Thus, on the morning of a blue and white day about the middle of the following July, astonishment blended with delight when the prolonged "Wee — , woo — ," echoed again in the old locality. I had kept a close census on that bit of bushland throughout the months, but never before had heard the Gilbert Whistlers in the district so early in the Springtime as this. It was an honor to be acknowledged.

On almost every day of the week following I visited the locality in search of the birds, but did not hear them calling again until early in August. Then the female, who showed no particular timidity, was feeding among the leaves of the trees, while her buff-throated consort, much more wary in manner, kept about the litter of leaves and bark on the ground. They remained constant to the same tract of timber during succeeding months, but were more often heard than seen; and many hours of searching failed to reveal a nest. The siren voices departed with the spring, but echoed once again about the favored spot in August of 1915.

That occasion was a re-union and a farewell in one; at a distance of a thousand miles from those quiet places, only memory knows now the lyric melody of the Gilbert Whistler:

> Sweet bird, that shunn'st the noise of folly,
> Most musical, most melancholy!

Chapter X

THE SPIRIT OF AUSTRALIA

(*A Study in Black and White*)

Consider what a potent element are black-and-white birds in the out-door life of Australia. It would probably be safe to say that we have more of these spruce studies, in numbers if not in point of variety, than any other country in the world. Indeed, the very familiarity of pied plumage tends to cause us to accept the colors and their owners as a matter of course, and to overlook the fact that, without them, there would be much less brightness and good cheer in Australia.

This conviction came to me with added force when travelling, in 1918, with the French Mission in Queensland. The soldier-visitors saw a good deal of the country areas, and, keen observers for the most part, they could not help noticing that there were almost as many wayside Wagtails, Pee-wees, Magpies, and pied Butcher-Birds to greet them as there were school children — "and all as happy." Maybe, then, these vivacious birds had something to do with the appreciation aroused in Commandant d'Andre, who, as we swept across the rolling miles of the Darling

Downs, proclaimed, with an expressive wave of the hand, his joy in the *couleur* of the country. "I understand," he said — this soldier with the heart of a boy and the soul of an artist — "why your soldiers are so merry. I understand the Spirit of Australia!"

The Spirit of Australia! Hear E. S. Emerson, who, more than all others, has put Australian birds into rhyme; analyse it in both Anzac man and bird: —

> Yet I've often thought, when resting, how this fighting
> spirit flashed
> Into mastership within one fleeting year;
> And I've wondered, as my comrades into war-old
> vet'rans dashed,
> How it was they never showed a sign of fear.
> But the riddle is no riddle as my thoughts the distance
> span,
> And in memory the mountain-track I take,
> For I've seen a nesting Magpie swoop undaunted on
> a man;
> I have watched the 'Burra kill a tiger snake.

The poet has seen, too, evidence of the pluck of the Wagtail, the Pee-wee, the Swan, and many another native bird, and he concludes: —

> That's the spirt of Australia. Never mind how great
> the odds;
> Never mind how long and bitter is the strife —
> To the death the bush-birds wage it; and, by all the
> living gods,
> That's the spirit that our menace brought to life . . .

THE SPIRIT OF AUSTRALIA

> Oh, my native land, down under, with your sunshine
>   and your song,
>   With the soul of all your bushlands for a throne,
> Though our heritage was freedom, and our fathers
>   bred us strong,
>   You have crowned us with a glory all your own.

Is there, one wonders, any particular reason why our black-and-white perching birds, above all others, are a mixture of music and fight? On the fact itself all Australians will agree. (Indeed, there was a suggestion put forward a short time ago that the Magpie, erroneously so called in the beginning because of its superficial resemblance to the pied bird of England, should be re-christened the "Anzac-Bird.") It is true that there are one or two black-and-white birds of a non-fighting nature, but it will usually be found that these species are dwellers in the forest or scrub, that they are less plentiful than the pied birds of the open spaces, and that in practically every case, the females of the species (as distinct from the wives of such birds as the Pee-wee, the Wagtail, and the Magpie) are quietly garbed.

Carrying on the reflection, it seems to me that so many common Australian birds are not fightable in spite of their conspicuous coloring, but *because* of that very fact. They have none of the protective coloration which plays so large a part in the scheme and balance of Nature; therefore they must needs be endowed with some other special ability to protect themselves and their offspring. Further, as that same conspicuous coloration would render negative any general attempt on the part of such birds to obscure themselves, Nature has decreed that they shall face the world boldly — *protected by their very prominence.**

Support is added this supposition when one considers the ways of such birds as the Black-fronted Dottrel and White-fronted Chat. Each of these ground-dwellers is boldly marked with black and white on the breast, while the back is colored in drab harmony with the habitat of the bird. It is, beyond all doubt, this factor of protective coloration that each bird relies upon for safety; the boldly marked breast is kept carefully turned away from the observer, and the obscuring effect of the duller plumage is really wonderful. You get the same cause and effect in the case of the old-world Lapwing, which, by the way, is the original "Peewit," or "Peewee," names commonly applied, by reason of a similarity in calls (rather than in color and flight), to our black-and-white Magpie-Lark.

Moreover, while black-and-white-fronted birds of the ground lack the straight-out fighting ability of the more prominent pied species, Nature has given them (in addition to the modicum of protective coloration in respect of both plumage and eggs) the compensatory power to avert danger from their nests by means of a lure. Who that has met the little "Nun" (Chat) at its nest will forget the splendid artistry of the pretty creature in feigning to be injured? At the first sign of possible danger to eggs or young, they — both sexes practise the ruse, and sometimes several birds join forces — will go fluttering and tumbling along in most realistic fashion, meanwhile uttering cries pitiful enough to suggest the

---

* "It is interesting to note," writes Mr. H. J. Massingham, in his diary of a pied day spent among the birds of Southwest Dorset, "how kindred are the characters of mapgie and wagtail. . . . But the magpie, like the wagtail, possesses that impulsive, irrepressible temper which, alas! so often betrays him to the killer." — *Contemporary Review*, August, 1919.

THE SPIRIT OF AUSTRALIA

throes of death. Was this device evolved for confusion of the children of men? If so, it would seem that certain birds regarded man as quite a gullible creature, and that the fighting species considered themselves more than a match for him in a "scrap." Manifestly, however, human-kind was not considered in the arrangements at all; the pluck of some birds, and the cleverness of others, came into being from natural causes, and, with the advance of civilisation, they have blended into the national life out o' doors as part of the indefinable Spirit of Australia.

It is not only that our black-and-white birds are prominent in particular areas. More than any other birds of a definite color scheme, they are freely distributed over the whole continent. The Magpie, the Magpie-Lark, the Black and White Fantail (Wagtail), the Restless Flycatcher, the Black-throated Butcher-Bird, and the White-shouldered Lalage — each one of these is to be found in every one of the mainland States. It is the more retiring of the bi-colored species, such as the Whip-Bird, the Black and White Swallow, the Black and White Robin, and the rare Pied Honeyeater — birds without any assertive system or means of protection — that are restricted to certain favorable areas.

But the question of distribution becomes much more problematical when it is noted how few pied birds are known to Tasmania. Of all the birds mentioned above, only one (the Magpie) is a distinct resident of the island State, while, among the others, the Magpie-Lark and Lalage are recorded merely as accidental visitors. Why should this be? How is it that the common little Wagtail, for instance, is not known to Tasmania? Are it and the other missing "black and whites" of recent birth — subsequent to the formation of Bass Strait? Here is an interesting problem for the geographical student.

On the other hand, there is probably no area where pied birds are so prominent as they are in Southern Queensland. Brisbane, in fact, might well be known (in an ornithological aspect) as the City of Black and White. The plenteousness of water about the city and its environs, added to the somewhat humid atmosphere, makes for a plentiful supply of insect life, and, as a natural corollary, both the Wagtail and the Peewee have become almost domesticated — members of the family. It is much the same in other centres right along the Queensland coast.

"Magpie-Larks often feed with our fowls," writes a friend in the North, "and I was amused to note that some of them have acquired from the fowls the habit of scratching for food. They go round and round, walking backwards, and drag their feet clumsily through the dust."

The beautiful Restless Flycatcher, which closely resembles the Wagtail in appearance, but has singular little ways and habits of its own, is only an occasional visitor to the towns. So, too, is the Lalage; but, while the "Grinder" comes only singly or in pairs, the former capricious bird moves in scattered companies, and may only be looked for in the Spring-time. Scarcely less common are the Wagtail and Peewee in the thinly timbered forest and downs country of these northern parts, where the color scheme is stressed strongly by the addition of very many black-and-white Crow-Shrikes — Magpies and Butcher-Birds. This particular Butcher-Bird (the black-throated species) is much more plentiful in these areas than it is in Southern Australia, where the grey Butcher-Bird ("Whistling Jack") is the chief representative of the genus.

In regard to the Magpie, how interesting it would be to know why the well-known white-backed species of the south does not occur in the north — having in mind the fact that the Black-backed Magpie is common both to north and south. One wonders whether it is in spite or because of this fact that Mr. Robert Hall (whose useful bird books are well known) has laid down in an ornithological magazine his belief that the Black-backed Magpie is merely a sub-species, or off-shoot, of the white-backed bird.

In "certain areas," he says, "the sub-species is so fixed as apparently to be a species; in others the inter-breeding and the specimens showing reversion are so common as to make them inseparable; while, again, in the back country of all the eastern States, is shown the strong evidence of lesser dimensions, apart from dichromatism. Yet these lesser dimensions are not quite confined to the inland and drier area. Their points of resemblance are so many and those of difference so few that one strongly inclines to mark them as one variable species. In habitat both are the same; flight, gait, mode of hunting for food, and the food itself are the same. . . . . The difference appears to be in the plumage markings; possibly, too, in warble and temperament, varying with the area. . . . The warble of the Black-back consists of about 12 distinct syllables, and is finished with an indescribable, delightful, jubilant note. In every instance the female commences the warble, the male falling in at the last note, but holding it out longer than the female. The White-back rarely indulges in a song, rendering it in less musical style, and only in a chorus. The syllables are fewer and shorter. . . . In temperament the White-back appears to be the more savage of the two. In breeding-time the White-back will attack almost any living thing of large dimensions, the Black-back rarely interfering with anybody. . . ."

Irrespective of whether bush dwellers in Southern Australia will agree with Mr. Hall's comparative criticism of the carol of the White-backed Magpie, there is no room for dispute as to its cheerfully truculent nature. When a pair of these masterful Australians "peg out a claim" in the Spring-time they assume a monarchy over all they survey — and there is a special embargo, born of experience, against roving boys.* The policy is even more offensive than defensive; swishing wings may hum about the ears of innocent youth ere he is within a quarter-mile of the sacred nest. Nerve-racking in its suddenness, this rushing sound is calculated to upset the equilibrium of a bush-loiterer who, engrossed in other interests, has failed to catch the initial war-cry of the bird. Full many a time both warning trump and beating wings have hastened my laggard footsteps along the bush tracks to school.

There is, I gather, a superstition still current in various parts of rural Britain to the effect that it is unlucky to see Magpies under certain conditions, these varying importantly with various localities. Thus, in some counties two are said to bring sorrow, in others joy; while in some places, we are instructed, one Magpie is a signal of misfortune, which can, however, be obviated by pulling off your hat and making a polite bow to the judicial bird. In Lancashire they say:

> One for anger,
>   Two for mirth,
>   Three for a wedding,
>   Four for a birth;

---

* From a small schoolgirl's essay: "I am trying to write on the birds, but my selfish brother has taken all the best things I was going to say. I don't call that fair, and I am going to laugh the next time the Magpies take his hat and peck his head when we are going to school. They don't like boys, and neither do I."

> Five for rich,
> > Six for poor;
> > Seven for a witch,
> > > I can tell you no more!

If only for the sake of Imperial propriety, it should be emphasised that the British Magpie and its Australian namesake, birds with title and color in common, are different, if not distinct, in disposition. The general affinity of black-and-white birds has been shown to hold good in an international sense, but the outlook of the Australian bird is freer, more spacious — its disposition more resolute. (The advanced democracy of the land is not confined to the genus *Homo*). To pull off one's hat to a sentinel "Anzac-Bird" would certainly be construed into a breach of the peace; and, moreover, closer approach to the nest would render handling of the hat distinctly unnecessary — the strong bill of the bird would then be in action. Verily, the late Frank Myers, devotee of Australia Felix, had much justification for writing of Mag., his "happy, dear, and blessed bird," as "unlike any other creature unto whom was ever breathed the breath of song ... unexampled as the forest he overlooks."

So there you have it — cheerfulness, courage, and originality! Setting aside other attributes of this essentially Australian bird, in these qualities lies the reason for the appreciative, if impracticable, suggestion that the Magpie's borrowed name should be altered to that of "Anzac-Bird."

The Black-throated Butcher-Bird has much in affinity with its relative. Though slightly smaller than the Magpie, it has the same build and general color-scheme, inhabits the same class of open country, and is possessed also of pronounced musical and fighting ability. I have heard the

Magpies singing their love-songs to the morn (and also to the moon) on many occasions, but not even those carolling choruses were sweeter than the concert performance of a quartette of Butcher-Birds rollicking in the bush beside Moreton Bay on a day in October. The notes most freely heard from individual birds lack the continuity of the Grey Butcher-Bird's spirited melody, and constitute a chortle rather than a song. But they are indescribably rich and pure, and add a certain gipsy wildness to what is probably the nearest approach made by any Australian bird to the full, rolling tones of a pipe organ.

Like its grey relative — and seemingly in despite of the solemn appearance of the genus — the pied Butcher-Bird sometimes becomes as merry on the wing as any Skylark — more so, probably, for the unceasing individual efforts of the little European songster are apt to bring it under the suspicion of being a trifle mechanical. And there was emphatically no trace of this element in the movements of a quartette of Butcher-Birds which greeted a Christmas morning near the border highlands of Queensland and New South Wales. Joying in each other's company and the freshness of the morning, they left the trees and swung up and down in the air, both movements and voices having the breezy freedom of merry children.

One other memory of the mellifluous voice of this gifted bird belongs to a sub-tropical mountain. Sufficient jungle had been cleared to meet the requirements of the sun-loving Butcher-Birds, and on a vital morning of September two of them came into a dawn symphony reminiscent of John Ford Paterson's picture in the Melbourne Art Gallery. Here, however, was a living picture; nor was it the less harmonious for the marks of man's presence. A massed array of white Watsonias was flowering radiantly at the edge of the jungle-hemmed paddock, and as the deliberate notes of the pure-voiced birds rose and fell, how perfectly the music seemed to harmonise with the *color* of the white flowers! It is doubtful whether any bird-voice can be out of symphony at sunrise on a bright morning. There is always the lyric touch then, be the voice ever so harsh. But the melody of these Butcher-Birds got beyond the lyrical, developed a Miltonic fulness, as it were, and merged itself in the challenging purity of the white flowers. Anon, as the sun rose and an element of the garish intruded into the symphony, the notes of the birds took a slightly *golden* tint, this time approximating to the color of a rich rose that festooned a gateway hard by.

Of the pluck of the Butcher-Bird in defence of the sanctity of its home many instances could be cited; one ready to hand concerns the warrior whose picture appears in this book, and who lived in the north of Queensland, not far from Mackay. "He was moderately tame in the non-breeding season," writes Mr. W. G. Harvey, in sending the photograph, "and often came in search of scraps to the field where we would be eating our lunch. But when the nesting season commenced his whole nature changed. Whenever we had occasion to pass through his locality he would meet us before we were within one hundred yards of the nest, and persecute us right past it until we arrived at a point where a pair of Black and White Fantails had a nest in company with a pair of Magpie-Larks. These plucky birds always chased the Butcher-Bird back, and we were glad to be rid of him for a while. Curiously enough, the pet aversion of this Butcher-Bird was a little black-and-white terrier which used to follow us about. The bird always endeavored to peck the dog's ears, and during the incubation season they were more often bleeding than not. One afternoon we had the camera with us, and, after focussing on a branch, tied the unfortunate dog underneath. Before long, down came the bird — and the photo was taken!"

An interesting recital this — noteworthy not only for its testimony to the courage of various pied birds, but for the record of the Butcher-Bird's quaint antipathy to a dog with the same color markings as itself, and, moreover, for the illustration of the resourceful shrewdness of the enthusiastic photographer of wild birds.

Apparently the female Butcher-Bird is not so bellicose as her lord, but quite approves of his warlike attitude. At Roma (south-west Queensland) I saw a fine male Black-throat furiously attacking his reflection in a window in the heart of the town, what time his mate sat by and supplied a little cheering music. Unprotected glass would never have withstood the hammering of such a strong beak.

The joining of forces by the Magpie-Lark and Wagtail, as mentioned by Mr. Harvey, is not at all uncommon; nor is the practice exactly unique. In the far north of Queensland pioneering ornithologists find the nest of the remarkable Manucode, or Trumpet-Bird, consistently in proximity to that of the robust Black Butcher-Bird, the distance between the sites usually being about fifty yards. In this case, however, it would seem that the arrangement is not mutual, but that the shining "Bird of the Gods" deliberately seeks the neutral protection afforded by the war-like Butcher-Bird, without rendering any service in return. Further, a Manucode has been seen to drive away its "guardian" when the latter wandered too close to the aerial home of the greenish-black bird.

No such misunderstandings occur between the Wagtail and the Peewee. Quite frequently, in fact, and for year after year, they nest in the same tree, the finely modelled mud nest, as shapely as its graceful owners, occupying a position on a lofty, horizontal branch, and the beautiful little cup-shaped, cob-webbed home of the Wagtail resting unobtrusively on a lower limb. That there is more than chance in this arrangement may readily be believed; but one has to remember that other factors than the protective instinct favor the practice. The two species, for instance, though occupying distinct places in the scheme of natural balance, frequently feed in company on grassy flats. The larger bird has not the flitting Fly-catcher's ability to snap up insects disturbed by the warm breath of animals — I once counted 23 Wagtails dancing attendance on 17 cows in a Brisbane suburb — but it does much towards freeing them from pestiferous insects. Like the Wagtail of James Thomas's fine poem, the dainty Peewee may be seen:

> In its suit of white and black
>   On some old, sedate cow's back;
> Stopping now, as though to say:
>   "If I'm heavy, tell me, pray."

Be it noted, too, that the Peewee has almost as much to say for itself as the "merry, babbling, restless bird," whose "sweet pretty creature" (voiced by night as well as by day) is known to almost every Australian child. "What both birds lack in strength," observes Mr. Harvey, "they make up in noise." "It is called the Magpie-*Lark*," a small girl gravely informed me, "because it is so merry."

That interesting observation has much in accord with those of the youthful Queenslanders who announced their respective beliefs that the Wagtail keeps its balancing fan incessantly moving "because it is so happy," "to keep itself cool," and (a very small boy offered this suggestion) "to brush the skeeters away!"

Passing now from a very distinguished quartette of Australians — Magpie to Butcher-Bird, Magpie-Lark to Fantail — we come naturally to a notable exception to what I have ventured to suggest is a law under which ubiquitous birds colored black and white in both sexes face the world with boldness of voice and manner. The Restless Flycatcher, though frequently confused with the Wagtail, is slimmer in build, more glossy in the black of its back, and clearly distinguishable by a white throat. The female is similarly colored, with a touch of buff added to the chest. Despite its distinctive uniform, however, the Restless Flycatcher is neither so voiceful nor so fightable as its popular little relative. Nor is it nearly so numerous — a fact that may be due to the strange lack of black-and-white assertiveness or to the bird's comparative weakness in nest construction. I knew a Victorian instance in which a pair of these birds, after scratching feebly about on a dry limb, gave up the attempt at nest-building, whereupon a pair of Wagtails rapidly put together a nest, of similar material, in the abandoned position.

Let it be observed, also, that practically all of these bi-colored bird acquaintances of ours are stay-at-home species. Seldom do any of them wander far from one locality at any season of the year. Withal, it is recorded by a friend who carried out a long droving trip down the west of Queensland, that a "Shepherd's Companion" followed his cattle for at least 70 miles, occupying ten days; identification was possible because the Wagtail had met with an accident and had no tail to wag!

The exception to the general lack of travellers among the black and whites is the White-shouldered Caterpillar-eater (*Lalage*), which cumbersome name, by the way, certain schoolboys cheerfully disregard in favor of the more expressive and alliterative title of "Midget-Magpie." Here, again, you get a courageous temperament, allied, in the case of the male bird, to a highly musical voice, even though the quiet garb of the female, and the very small size of the nest, afford protection that is lacking in the cases of so many other daring studies in black and white.

Chapter XI

THE PARADISE PARROT TRAGEDY

Of how many small birds native to Australia has it to be written in sorrow, "The beautiful is vanished and returns not?"

It is time we looked into this matter. It is time we gave over the self-centred idea that the spread of settlement necessarily means the extermination or serious decimation of the shyer native birds. It is time, too, that a national endeavor was made to save the residuum of certain fine Australian birds that are trembling on the verge of nothingness.

For over one hundred years, now, a new and virile human race has been forcing itself into the domain of the native birds of this country, and it would seem that we are undergoing, in this respect, a process that other civilised countries passed through hundreds of years ago, and of which little record has been left. Australia is, in the words of the poet O'Dowd, the "last sea-thing dredged by sailor Time from Space," and here we have an absorbingly interesting opportunity of observing how wild creatures become reconciled to the civilisation of the white man.

Already many of our more familiar birds are becoming semi-domesticated. One is almost persuaded to say that they have become "Anglicized." In the course of "rubbing shoulders" with man in thickly settled Britain, the birds of that country — probably of Europe generally — have become part of the human scheme of things. They have made themselves so. At all events, birds generally are more ready than men to appreciate and reciprocate good feeling.

A writer in the London *Times*, recently, laid it down that "wild" birds were all originally inquisitive and friendly with man, "as the penguins of the Antarctic were when he first met them." Probably so. And how narrow is the line between bird behavior as it was in the beginning and as it is at the present day under favorable circumstances! Viscount Grey has lately pointed out (an illuminating column in the Manchester *Guardian* gives his views on the national importance of birds) that it is easy to create sanctuaries where birds "remain wild but lose their fear of man." We have had ample evidence on the point in Australia.

Not all wild birds are on the way to becoming tame — what a mournful prospect if that were so! — but of a surety many kinds are holding out, as it were, the wing of friendship, and maintaining a guarded amity towards Nature's most dominant production. There can be little doubt that if everyone in Australia could be persuaded to act in brotherly fashion towards native birds for a few years, the story of St. Francis of Assisi would pale by comparison with our experiences of every day. But in some cases, alas! even were this starry ideal put into practice, immediately, it would come too late.

In this respect I think chiefly of several of our beautiful Parrots. The history of these birds constitutes something in the nature of a national tragedy. Consciously or not, the

THE PARADISE PARROT TRAGEDY 155

white man in Australia has been progressing to a sad tune in the ears of a naturalist — to the dirge of other natives besides "the old people." He has destroyed, unnecessarily, the feeding-grounds of valuable birds; he has caused them to be captured in thousands and held for his own trivial entertainment; and he has carelessly foisted so many enemies upon them (the fox, for instance) that many birds have simply withered away, not as individuals, but as species. For example, consider the case of the regal Queensland bird known in Europe of old as the Paradise Parrot, and called by people of its native country the Beautiful or Elegant Parrot.

It was nearly eighty years ago that John Gilbert, able coadjutor of the great John Gould, the "father" of Australian ornithology, when carrying out ornithological work on the then recently discovered Darling Downs, shot a Parrot of a species he had not previously seen. Gould referred the specimens to the genus *Psephotus*, and, filled with admiration of the beauty of the birds, gave them the specific title of *pulcherrimus*. "The graceful form of this Parakeet," wrote Gould, "combined with the extreme brilliancy of its plumage, renders it one of the most lovely of the *Psittacidæ* yet discovered; and in whatever light we regard it, whether as a beautiful ornament to our cabinets or a desirable addition to our aviaries, it is still an object of no ordinary interest."

Superlatives having been wrung from a seasoned scientist, who saw only lifeless specimens of the "most lovely" bird, what was to be expected from those persons fortunate enough to know it in life? But, strangely enough, little was written about the species until the 'eighties. By that time, apparently, large numbers of Gould's "beautiful" Parrot had been sent abroad for aviaries, and had become known to the bird-dealers of Britain and the Continent under the name of Paradise Parrot.

What a degree of popularity the shapely little Australian enjoyed (!) is made evident by Dr. W. T. Greene, M.A., in his finely illustrated book, *Parrots in Captivity*, published in London in 1884, and now little known in Australia. After describing the "Beautiful or Paradise Parrot" as more lovely, if possible, than the Many-colored Parrot, also of Australia, the writer says: "No one can see it without desiring to possess so beautiful and graceful a bird, and large sums are constantly being paid for handsome specimens by amateurs; but, alas! one in a dozen survives a few months and — dies suddenly in a fit one day."

Further, the Rev. F. G. Dutton, a correspondent of Greene's, improves on his colleague's tribute by saying roundly: "*Psephotus pulcherrimus*, the Paradise Paroquet, as the dealers call it, is not only the most beautiful *Psephotus*, as its name says, but surely the most beautiful Paroquet that exists. The vivid emerald green and carmine of the cock, beautifully contrasted with the grey of the rest of the plumage, make him 'a joy for ever.'"

Although nothing was known in England, then, concerning the Paradise Parrot's curious habit of nesting in termites' mounds, this practice was more or less familiar to certain Queensland settlers many years before the date of Greene's book. To them the bird was, variously, the Ground Parrot, Ground Rosella, Beautiful Parrot, Elegant Parrot, and Anthill Parrot, to which multitude of titles was added later the name of Scarlet-shouldered Parrot. In many districts it was a favorite cage-bird, though perhaps no more so than outside its own country.

The Barnard family, of Coomooboolaroo, near

THE PARADISE PARROT TRAGEDY 157

Rockhampton, were among the first people with ornithological leanings to take note of the nesting-habits of the "Ant-hill" Parrot. When Carl Lumholtz, the Norwegian author of *Among Cannibals*, was at Duaringa in 1881, he was introduced by the Barnard boys to the burrows of the beautiful bird in termites' mounds, and of these he penned an interesting description. On another occasion, near the Nogoa River, Lumholtz had an experience with a pair of these birds that deserves to be revived from the semi-obscurity of his book.

"An hour before sunset," he says, "I left the camp with my gun, and soon caught sight of a pair of these Parrots, male and female, that were walking near an ant-hill, eating grass-seed. After I had shot the male the female flew up into a neighboring tree. I did not go at once to pick up the dead bird — the fine scarlet feathers of the lower part of its belly, which shone in the rays of the setting sun, could easily be seen in the distance. Soon after, the female came flying down to her dead mate. With her beak she repeatedly lifted the dead head up from the ground, and walked to and fro over the body, as though to bring it to life again; then she flew away, but immediately returned with some fine straws of grass in her beak, and laid them before the dead bird, evidently for the purpose of getting him to eat the seed. As this, too, was in vain, she began again to raise her mate's head and to trample on the body, and finally flew away to a tree just as darkness was coming on. I approached the tree, and a shot put an end to the faithful animal's sorrow."

That little tragedy will serve, fittingly enough, as an introduction to a dark chapter in the history of the species generally. Possibly that sad phase had its genesis much earlier, with the establishment and spread of settlement.

Howbeit, the fact is that as the years went by the Paradise Parrots steadily decreased in numbers. In time they became an unknown quantity on the markets overseas. In time, too, they vanished from districts where once they were a feature — a very beautiful feature — of the subtropical landscape.

The decimation attracted no particular attention in ornithological circles until 1915. Then Mr. A. J. Campbell, C.M.B.O.U., wrote in *The Emu* an article entitled "Missing Birds," specifying in this respect the Paradise or Scarlet-shouldered Parrot, the Turquoisine or Chestnut-shouldered Parrot, and the Night-Parrot. "It would be interesting to know," wrote Mr. Campbell, "if these three beautiful Australian Parrots still exist or have been exterminated. If the birds are extinct, what is the cause or causes of their extinction?"

There was no response to the inquiry. The Paradise Parrot, it appeared, had been lost in annihilation's waste.

So it seemed to ornithologists in Australia, and so it seemed to Mr. Gregory Mathews in England. "It is a matter for deep regret," he wrote in *The Birds of Australia* (1917), "that this most beautiful of Parrots appears to have become extinct without any lasting record of its life-history being made known." Further, in referring to another Parrot, not yet uncommon, Mr. Mathews advised study "before it becomes extinct, like its congener, *P. pulcherrimus.*"

That was the position when, in the middle of 1918, the subject was taken up afresh in Queensland, the former stronghold(!) of the missing bird. Hints gathered in conversation with old settlers had indicated that further search would be at least worth while. Accordingly, letters on the point, bearing the query-caption, "Is It Lost?" were directed to and published by the leading daily newspapers

THE PARADISE PARROT TRAGEDY

of Brisbane and the Darling Downs. The response was prompt. It was also partially satisfactory. Most of the replies dealt with the species only from a posthumous viewpoint, but, again, there were hints that served to strengthen the belief that odd members — "the ever-blessed residuum" — of the beautiful species might yet be found.

Well, for three years the benevolent pursuit of the "lost" Parrot was continued intermittently. And intermittently there continued to float in suggestions and whispers regarding the existence and whereabouts of odd members of the species. Occasionally, something more definite arrived. For instance, a bushman living in an out-of-the-way spot between Bundaberg and Gladstone reported in 1919 that the missing "Red-shoulder" was to be seen about his locality. He knew nought of its distinctive breeding habits, but mentioned that some of the birds could be seen in captivity.

Accordingly, Mr. C. T. White, Queensland Government Botanist, and I took train on a night in April of 1920, travelled 250 miles in that manner, walked ten miles through inhospitable country, and crossed a broad creek in a leaky boat, what time hordes of ravening sand flies scored our bare legs — all to find that the local Parrot was the common Red-wing (*Ptistes*), a bird that could be seen in a dozen cages half a mile from our homes! The irritating similarity of vernacular names had deceived our friend.

Still, as old John Burroughs once said, "Whichever way I go I am glad I came;" among other arresting sights of the locality were a pair of White-eared Flycatchers and two young, this being probably the "farthest South" record in the breeding range of that rare and little-known bird.

We come now, somewhat belatedly, to more recent and thoroughly definite developments in the search for the Paradise Parrot. On December 11th, 1921, Mr. C. H. H. Jerrard, a keen naturalist and capable photographer, wrote from the Burnett country (Queensland) to say that he had seen a pair of Parrots which he was almost sure were *Psephotus pulcherrimus*. A description which he supplied, and which fitted the species, was made out as the birds perched in a tree, but for portion of the time when watched they were on the ground. Less than a week later Mr. Jerrard became sure of his birds, having his opinion reinforced by a neighbor who had kept the "Ground Parrots" in captivity many years before.

Here, at last, was a report that was not only definite, but was made by a man who was competent to follow it up. His attention having been directed to the termites' mounds, Mr. Jerrard soon found holes suggesting the breeding-hollows of the Parrots. In more than one case there were signs that nesting operations had been commenced and then left off. But the year drew to its close without any discovery of an actual nest, and the scant literature on the subject having given September–December as the breeding period, there remained but little hope of a pair of the beautiful birds being studied "at home" for many months.

Sub-tropical birds, however, swayed by a willful climate, are not as other birds are in the matter of breeding seasons. So, it was not altogether surprising that the patient watcher was able to report, on January 21st, 1922, that a pair of the Parrots had recommenced work on a hollow that had previously been visited. On that date Mr. Jerrard watched the mound for two hours, but was chary of investigating closely. He saw the male bird pay one visit of inspection, without actually entering the hollow, the female being in the vicinity at the same time. That caution of the observer was wise. Subsequent observations made it clear that eggs were not laid in January, for it was not until March 4th that Mr. Jerrard was able to report with certainty that the female Parrot was brooding.

From that point the watcher waxed keener than ever in his fraternal spying upon the rare and lovely Parrots. Working with care, he erected a rough hessian shelter in front of the exposed little hillock that afforded the birds a home, and from this vantage-point was able both to study and photograph the pair. On March 18th, 1922, Mr. Jerrard sent me the first picture ever taken of the Paradise Parrot at its nest. This photograph depicted both male and female, and showed the regal little head of the house to be the bolder bird of the two.

Indeed, though apparently the female was sole custodian of the eggs — the male was never seen to enter the tunnel — she was much more nervous than her mate. Frequently he would accompany her to the hollow, which she would at once enter and remain in for periods extending from half an hour to two hours. The ceremony attached to her re-emergence was both interesting and pretty. The male would alight on the mound, and, looking into the hole, emit soft, sweet chirps until the faithful little home-keeper answered by coming out and flying off with him.

Is not this practice in affinity with the methods adopted in the conduct of the homes of Hornbills? The male Paradise Parrot is evidently master of his own household, and were he not an entirely amiable bird, as old English aviculturists assure us was found to be the case, he might have developed — who knows? — the domineering tactics of the Hornbills, and walled his mate in the nesting-hollow for the term of her breeding period. But this is rather an idle supposition to apply to a *menage* which is obviously ruled by affection. Certainly, the regal bird would seem to take to himself more freedom and ease than his sober little consort enjoys; but who will say that all this grace and beauty should be hidden away in a dark hollow at any time?

Further, Scarlet-shoulder is apparently the melodist of the pair. "He has a musical and very animated song," writes Mr. Jerrard. "I heard it in October of 1921, and noted how his whole body vibrated with the force and intensity of his musical effort, imparting an agitated motion to the long tail, which bore adequate testimony to the vim of the performance. It all seemed to indicate a very intense little personality under the beautiful exterior."

Considering all the circumstances attached to the species, what would any reader of this story have aimed at in the case under review, apart from placing on printed and pictorial record something of the life-history of the species? We thought the matter over, and came to the conclusion that it would be best to take some of the young from that nest in the public interest. It might be possible to have them breed under authoritative control; but at least thousands of people who would wish to see live specimens of a distinguished Queensland bird should be given the opportunity to do so — under proper conditions. Alas! that amiable scheme was doomed to failure. On April 8th Mr. Jerrard reported that some mischance had intervened to prevent the eggs being hatched. He had reason to believe that incubation had commenced before the beginning of March, but, judging by the behavior of the birds, there was no indication of young near the end of that month.

Any further history attaching to that nest can be told briefly.

"In accordance with the suggestion contained in your

last letter," wrote Mr. Jerrard early in May, "I opened up the nest on April 24th, there being no longer any doubt that it had been deserted. The enclosed photographs show the result of that investigation. I was careful not to touch the eggs before photographing them. They had not been disturbed, but seem to be all addled. One was punctured and the contents dried up; another I broke, and found it to contain nothing but stinking fluid. No embryo seems to have developed in any of them."

Then follow these notes, taken on opening the mound: "The entrance tunnel is about nine inches long and one and a half inches in diameter. It enters the nesting-chamber at the top and to one side, so that the eggs cannot be seen or touched from outside. The nesting-cavity is roughly circular, about 15 to 18 inches in diameter and eight inches high in the middle. The light, honeycomb material in which it is excavated had not been carried outside (as in the case of the harder material through which the tunnel is bored), but lies at the bottom, forming the bed of the nest, on which the eggs lie. There is no other material whatever. The floor of the nest is lower than the ground outside. The eggs, five in number, are white, with a pinkish tinge, and measure .9 in. x .8 in. Both ends are shaped nearly alike. They rest under the centre of the mound. There were no termites in the mound when I opened it."

No further nests of the kingly Parrot have come under notice, but from time to time, both before and after the discovery of the historic nest, Mr. Jerrard's keenness has rewarded him with the sight of one or two other families of the Paradise Parrots; the quiet district, it would seem, is one of the few and final refuges of the brilliant birds. Attention was usually called to the Parrots by the short and sharp but musical whistle uttered before taking flight from the ground on the approach of danger. Mr. Jerrard thinks that the old birds are constant to one nesting locality, year after year, and that some of the offspring subsequently nest in proximity to the parental home. He has never seen one of the Parrots more than a mile from the spot where he first discovered them in 1921.

The general observation recorded in the two preceding sentences suggests more than meets the eye. It gives rise to a suspicion as to why the eggs in the nest under review failed to hatch out. May not the fact be that too close breeding, consequent upon the extreme scarcity of these Parrots, impaired the fertility of the eggs? This deduction appeals to me as feasible, and it strengthens by what the observer has to say, retrospectively, regarding the pretty perplexity of the birds at the non-appearance of their expected family. Experience having taught Mr. Jerrard, as it has taught others, that undue familiarity may cause breeding birds to leave their eggs, he waited until he was reasonably sure that nestlings in the little hillock were well advanced. Then, to avoid any unnecessary "pottering" about the nest, he erected his cube-shaped tent at home, all in one piece, so that it was set up in a few minutes by slipping it down over four stakes, erected in position. Thus, the birds showed little, if any, resentment of the intrusion, and home-keeping proceeded smoothly.

Ultimately, however, there came a day when the male Paradise Parrot visited the mound with his usual confidence; but when he stood looking in the hole with one eye, Parrot-wise, and chirping softly — perhaps inquiringly — there was an almost comical suggestion of perplexity in his manner. This was heightened by the dejected look of his mate, who sat for some time on top of the mound, but did not enter, as she had always done

THE PARADISE PARROT TRAGEDY

before. It was quite plain that the kingly bird was sorely puzzled by the failure of his family hopes, and that the poor little wife, on her part, took the disappointment doubly to heart in that aspersions seemed to be cast on her maternal qualifications!

Here, then, a remarkable situation is suggested. It does not follow, from the observations and theories recorded in the two foregoing paragraphs, that the "stay-at-home" proclivities of the Paradise Parrots have been an original factor in reducing their numbers, but it does appear probable that the beautiful birds have been reduced to a point where natural increase has become warped — a point at which they are, as it were, *exterminating themselves!*

In the course of the search for the missing Parrot, it became evident that the name "Ground Parrot" was the most familiar one for the bird, "Elegant" that by which it was known to dealers, and "Ant-hill" Parrot the most definite title for identification purposes. It was, indeed, the bird's habit of nesting in termites' mounds, no less than its graceful and pretty ways, that made it noticeable in earlier days, the only other Parrot known to follow this practice being the closely allied Golden-shoulder, of the far North.

Incidentally, it is curious to reflect that the remarkable nesting trait of these two Parrots is shared by certain other species of birds possessed of long tails — a factor that would seem rather opposed to occupancy of an earthern burrow.* The beautiful Long-tailed Kingfisher, of Cape York, also breeds in termites' mounds. Further, *Merops ornatus*, the so-called Bee-eater, which is graced with two long, feathery shafts extending beyond the tail,

---

* The Paradise Parrot averages 13 inches in length, of which seven inches is tail.

always makes its nest by burrowing in a bank or in sandy ground with, preferably, a slight slope. Why this point of similarity between birds whose only other feature in common is the possession of long tails?

This aside, it cannot be doubted that the unusual nesting habit has had a good deal to do with the falling away of the Paradise Parrots. Correspondents unite in asserting that the "goanna" found them an easy prey, and that trappers secured them with a minimum of trouble. Considerations of food, affected through human agency, must also be looked to when reflecting on the disappearance of the species. Weakened by the competition of introduced domestic animals, the birds were in no condition to withstand Queensland's big drought of 1902.

Having in mind, then, the effect of trapping, the burning of grass, and the ravages of domestic cats gone wild, it seems moderately clear that the "most beautiful Parrot that exists" has been brought to the very verge of extinction by human agency, following on Nature's indiscretion in bestowing upon it the fatal gift of beauty without adequate means of defence or protection. It is all very lamentable. It is more; it is a national calamity. Both the citizens and governing authorities of Queensland have neglected a definite duty — a duty to helpless beauty — in allowing these pretty birds to be sacrificed. Whether it is too late to make amends cannot well be said; but the authorities showed the right spirit, while these inquiries were progressing, in extending full protection to the "lost" species and all kindred Parrots.

Finally, let us look for a few moments at the position of Australian Parrots generally. Mr. Mathews expresses the view (*Birds of Australia*, 1917) that these birds "have shown signs of extinction in a very rapid manner." I agree with

THE PARADISE PARROT TRAGEDY

him. It seems to me that the Parrots of the mountains, the King and Crimson species, for instance, are holding out fairly well. So also are several of the broad-tailed Rosellas. Among the Grass-Parrots the little Budgerigah and the Red-backed Parrot are still fairly common. Aside from these two latter species, however, there is not one of the Grass or Ground Parrots that has not "slipped" very seriously. Where now is the Night Parrot? How rarely the Green Ground Parrot is reported! What has become of the regal *Euphema splendida*, the Scarlet-chested Grass Parrot? (Once a plentiful species, only one company has been recorded in recent years, and that a small lot in South Australia.)

And what of that Paradise Parrot in miniature, the Turquoisine or Chestnut-shouldered Parrot? The late A. J. North, who had a wide circle of correspondents, wrote in 1911 that he had for years received inquiries from aviculturists concerning this bird, but had not been able to afford them any information, the last specimen received at the Australian Museum (Sydney) being dated 1886. (This of a bird which John Gould had found quite common in N.S.W. in the 'forties, and which another early birdman, writing in the 'sixties, alluded to as "this beautiful but common species!") Two years later (in 1913), Mr. W. H. Workman, M.B.O.U., wrote to *The Emu*, from Dublin, drawing attention to "the disappearance from the bird-markets during the last twenty years of the beautiful little Turquoisine Parrakeet," and expressing the fear that the species had "gone the way of the Dodo and the Passenger Pigeon."

"If our worst fears are realized," added Mr. Workman, "and this little bird has gone for ever, I think it would be of interest to ornithologists all the world over if a short history of the species were published in *The Emu*."

The editors of *The Emu* appraised the question as an important one, and asked ornithologists throughout Australia for notes upon the species, either from past or present experience. There was no response. Two years later appeared Mr. A. J. Campbell's inquiry ("Missing Birds"), to which allusion has been made earlier. Again there was no response. All this caused Mr. Mathews to write in his big work that the Chestnut-shouldered Parrot was probably extinct, "and of its life history we do not know much." Since then (in 1921, I think) a small company of the Turquoisine Parrots was reported not far from Sydney. I have not heard, however, of any attempt being made to follow out Workman's suggestion in regard to fostering the breeding of the birds.

The extinction of a species is a ghastly thing. How much more appalling is the extermination of a genus! Such a position confronts us in respect of the *Euphema* birds and other small Grass Parrots of Australia. Mr. W. B. Alexander, M.A., an ornithologist of wide experience, tells me he thinks Parrots are failing the world over; but he would be the last to admit that because of that belief we should sit down with folded hands. The idea that such birds must have their day and cease to be can well be left to the trappers and dealers, gentlemen who mix fatalism with finance.

The question arises, then, what are the bird-lovers of Australia going to do about this matter of vanishing Parrots? Surely it is a subject worthy of the closest attention of *all* good Australians! Meanwhile, let us, without reflecting on the claims of true science, dispute the dangerous idea that a thing of beauty is a joy for ever in a cage or a cabinet; and disdain, too, the lopsided belief that the moving finger of Civilisation must move on over the bodies of "the loveliest and the best" of Nature's children.

INDEX

Adams, Dr., on ornithology and education, 60
Alexander, W. B., on vanishing parrots, 168
Almond blossom, its beauty and appeal, 9
American bird names, 67
"Among Cannibals," quoted, 157
Anthill parrot, *see* paradise parrot
Anzac-bird, *see* magpie
Aristocratic birds, 87–103
"Around the Boree Log," quoted, 130
August, the appeal of, 3–15
Australia, the spirit of, 137; influence of its climate on poetry, 43; and on birds nesting, 45, 160
Australian parrots, in England, 155–6

Babblers, nesting in August, 14–15
Baby birds, their hearing at birth, 26; their incessant supplications, 44; imitating caterpillars, 50, 92–3
Banfield, E. J., on the sun-bird, 80
Bark-robin, *see* yellow robin
Barnard family, knowledge of paradise parrot, 156–7
Bayldon, Arthur (poet), on a bush canary, 130
Bell-miner, its strong claws, 86; its communistic chiming, 89 (*See also* crested bell-bird.)
Biglow Hosea, on the American spring, 31
Bird clubs, 60; bird play-parties, 83; bird tragedies, 85–6
Bird Day movement, 57
Birds, and civilisation, 153–4; their names discussed, 66–8, 106, 165
"Birds of Australia," quoted, 158, 166–7
Blackbird, English, in Australia, 34, 45
Black butcher-bird, alliance with manucode, 148
Blackfaced cuckoo-shrike, its notes and manners, 20; concert by, 21
Black-throated butcher-bird, its abundance in Queensland, 141; its organ-like voice, 146; singing while flying, 146–7; color of its song, 147; its strong pugnacity, 147–9
Black-throated warbler, imitating caterpillars, 93
Black-and-white birds, their prevalence in Australia, 137; reason of their combativeness, 139
Black-and-white robin, 115, 141
Black Swans, in poetry, 36, 138
Blood-bird, *see* scarlet honeyeater
Blue, in flowers, 11–13
Bluie, *see* wood-swallow
Boarded-out children, their love of Nature, 59
Bobolink, American, 67

INDEX

Bottlebrush, as host for birds, 78
Boys, as nest-robbers, 19, 27; their names for birds, 9, 20, 45, 90, 151
Brereton, J. Le Gay (poet), on the Jack o' Winter, 4
Britain, superstition in regarding magpies, 145
British birdlover, sighing for spring, 43, 54
Brisbane, its birds, 117; as a city of black and white, 142
Bronze cuckoo, as a spring visitor, 7; answering a call, 7; two eggs in a robins' nest, 112
Brown, Dr. John, on value of Nature-study, 59–60
Brown flycatcher, greeting the spring, 4; following a relative, 4
Budgerigah, its prevalence, 167
Burroughs, John, on birds' cleverness, 28; on an American thrush, 47; on simple names, 66–7; on a neighborly bird, 107; on the joys of rambling, 159
Bush-chats, *see* white-fronted chats
Butcher bird, *see* black, black-throated, and grey butcherbirds

Campbell, A. J., on decimation of parrots, 158, 168
Cardinal, *see* scarlet honeyeater
Caterpillar-eater, *see* white-shouldered caterpillar-eater
Caterpillars, ravaging forests. 50–1; in bell-birds' nests, 93; imitated by baby birds, 51, 92–3
Cats, ravages among birds, 166
Chickowee, in almond blossom, 9
Children, in Bird-land, 57–68; at a bush cemetery, 66; their attitude to technical names, 66; their quaint letters, 61–5; their love for robins, 106

Clarke, Marcus, on the "weird" bush, 23
Cockatoos, pride in crest, 87–8; their prevalence, 88
Cockatoo-parrot, its prevalence, 88
Cole, Elsie (poet), on voice of thrush, 49
Coutts, C. B. (poet), on voice of thrush, 49
Coloration, protective, 73, 122; of birds' eggs, 45, 86, 132
Crested bell-bird, its thrush-like nest, 26; nesting in dry weather 49; its shyness at the nest, 92, unfledged baby bird hops away, 49; its queer practice with caterpillars, 51, 91–3; its beautiful chiming, 89–91; contrasted with bell-miner, 89–90
Crested shrike-tit, appearance and habits, 93–103; its powerful piping, 8, 102; search for its nest, 94–5; its behavior at nest, 95–100; how the nest is built, 97; desertion of nests, 97–8; photography of, 99–100; quivering of wings, 100; visiting of towns, 101; its dignity and combativeness, 101–2
Crimson parrot, its prevalence, 167
Cuckoos, problem of their parasitism, 53–4; enormous appetite of young, 54 (*See also* bronze, fan-tailed and pallid cuckoos.)
Cuckoo-shrikes, their cuckoo-like flight, 20 (*See also* black-faced cuckoo-shrike.)

Daley, Victor (poet), on a memorable day, 63
D'Andre, Commandant, on the spirit of Australia, 137
Darling Downs, home of paradise parrot, 155

INDEX

Dennis, C. J. (poet), on the robin, 52; on the thrush, 49; on sunshine and bird-song, 82
Dick the devil, *see* crested bell-bird
Diuris orchids, 33, 66
Dunk Island, 80
Dutton, Rev. F. G., on Australian parrots, 156

Earth, "mad with song," 24; "heaven on earth," 55
Echong, *see* rufous-breasted whistler
Elves of spring, 22, 129
Emerson, E. S. (poet), on the spirit of Australia, 138
Eriostemon (waxflower), its daintiness, 13
Euphema parrots, 167–8

Fairyland arithmetic, 65
Fantail, *see* wagtail
Fantailed cuckoo, its trill, 20; its parasitism, 54
Fasciated honeyeater, 82
Finch, spotted-sided, 20
Flame-breasted robin, its showiness, 119; separation of sexes, 119; its curious movements, 119
Flowers, of almond tree, 9; of the bush, 11; in harmony with a bird song, 146–7
Fuscous honeyeater, nesting in flowering wattle, 14; young birds take fright, 50; its nesting, 85
French Mission in Queensland, 137
Fox, destroying birds, 155

Gilbert, John, discovery of Gilbert whistler, 132; discovery of paradise parrot, 155
Gilbert whistler, its color, nest and eggs, 132–4; its rarity, 132–3; its sweet voice, 133; its fondness for old haunts, 134
Gizzie, *see* little lorikeet
Golden-breasted whistler, its beauty of plumage and voice, 127–8; its distribution, 126; curious impulses and courtship, 128
Golden-shouldered parrot, its curious nesting habits, 165
Goldfinch, English, in Australia, 34, 45
Gordon, Adam Lindsay (poet), his delight in the Australian spring, 18; on "songless bright birds," 121
Gore-Jones, Alice (poet), lines quoted, 17
Gould, John, on movements of robins, 116; on a whistler's voice, 123; on the genus of whistlers, 124–5; on the paradise parrot, 155
Gould League of Birdlovers, 62
Graucalus, *see* cuckoo-shrike
Greene, Dr. W. T., on Australian parrots, 156
Greenfinch, English, in Australia, 45
Grey, Viscount, on tameness of birds, 154
Grey butcher-bird, "bluffed" by lorikeet, 10; its song, 145–6
Grey thrush, a frightened mother, 26; singing while nest-building, 26; pretty eggs, 27; unusual nesting, 27; voices of young, 47; half-sane shout, 47; robbing other birds, 48; in poetry, 48–9
Ground-thrush, feigning to be injured, 28–9, 111–12
Ground parrots, 165

Hall, Robert, on magpies, 143
Harvey, W. G., on a butcher-bird, 147–8; on the magpie-lark, 147

INDEX

Heath, as host for spinebills, 14

Henry, O., on harbingers of spring, 3; on "enchanted days," 79

Hill, G. F., on caterpillars in birds' nests, 93

Honeyeaters, in winter and spring, 4; their blithesomeness and beauty, 73–4; their prevalence and habits, 75–6; distinctions in size and nests, 82–6; as weather prophets, 85; influence of food on eggs, 86 (*See also* fasciated, fuscous, New Holland, regent, scarlet, spinebilled, white-naped, white-plumed, and yellow-tufted honeyeaters.)

Hornbills, curious nesting ways, 161–2

Hudson, W. H., on "largeness" of spring mood, 11; on blue flowers, 12

Jacky Winter, *see* brown flycatcher

Jerrard, C. H. H., on the paradise parrot, 160–4

Jerryang, *see* little lorikeet

Joy-bird, *see* rufous-breasted whistler

Kendall, H. C. (poet), on "bright yellow tresses," 13; on the echong, 131

Kingfisher, nesting in termites' mounds, 165

King parrot, 167

Kookaburra, 34

Lalage, *see* white-shouldered caterpillar-eater

Lapwing, feigning to be injured, 140–1 (*See also* cuckoo-shrike.)

Larks, as harbingers of spring, 22–4

Lorikeets, their speed and fondness for travel, 76 (*See also* little, musk, purple-crowned, rainbow, scaly-breast and swift lorikeets.)

London "Times," on birds' tameness, 154

Lumholtz, Carl, on a parrot's faithfulness, 157

Macgillivray, John, his first sun-bird, 80

Macgillivray, Dr. W. D., on birds imitating caterpillars, 92–3

Macpherson Range, as habitat for rose robins, 119

Magpies, in early spring, 6; their wild warble, 143; singing at night, 145–6; their combativeness, 144–5; distinction from British magpie, 145

Magpie-lark, its prominence and plentitude, 141; its semi-domestication, 142; alliance with wagtail, 147–8

Manucode, alliance with black butcher-bird, 148

Manchester "Guardian," quoted, 154

Many-colored parrot, 156

Masefield, John (poet), quoted, 59

Mathews, Gregory, on vanishing parrots, 158, 166

McCrae, Hugh (poet), quoted, 9, 34

Merops, its nesting habits, 165–6

Migratory birds, charm of their call, 22; manner in time of danger, 28–9

Mistletoe-bird, 24

Mouth, yellow robin clinging to, 115

Murdoch, Nina (poet), on bush flowers, 11

Myers, Frank, on the happiness of magpies, 145

Naturalists, searching for nests, 25; keeping "appointments," 30

INDEX

Nature, as a joyous playground, 68; its beacons, 114

Nature-study, its place in schools, 58–9

Neilson, J. Shaw (poet), line quoted, 12

Nesting, sport of, 19; discretion of birds, 40

Night Parrot, its rarity, 167

Nogoa River, haunt of paradise parrot, 157

North, A. J., on the Gilbert whistler, 132–3

November's heat, 44

Oasis, in the bush, 33–4

O'Dowd, Bernard (poet), 67; quoted, 153

Orchard, bush, its appeal in October, 33

Orchids, glossodia, appeal of scent and color, 12; diuris, 33, 66

Ornithologists' Union, 105

Paradise Parrot, its discovery, beauty, popularity as cage-bird, 154–6; disappearance, 157–8; search for its, 158–60; rediscovery of, 160; photography of, 161; failure of eggs, 162–3; reason for rarity, 163–6; plea for its preservation, 166

Pallid Cuckoo, as harbinger of spring, 7

Parrots, decimation of, 155, 166. (*See also* budgerigah, crimson, golden-shouldered, king, night, many-colored, paradise, red-backed, red-winged, rosella, scarlet-chested and turquoisine parrots, and lorikeets.)

"Parrots in Captivity," 156

Paterson, "Banjo" (poet), 36

Paterson, John Ford, artist, 146

Peewee, *see* magpie-lark

Persephone, approach of, 3–4

Photography, in tree-tops, 98; of baby birds, 100; of a crouching bird, 110; of whistlers, 131; of a fighting butcher-bird, 147–8; of paradise parrots, 161

Pink-breasted robin, 115

Psalmist of the dawn, *see* yellow robin

Purple-crowned lorikeet, its discretion near houses, 9–10; "bluffing" a butcher-bird, 10; its restricted distribution, 10, 76

Pyrenees Range, beauty in August, 14

Queensland, drought, 45, 166; habitat of paradise parrot, 156

Quinn, Roderic (poet), on the inspiration of birds, 67; on a poet and gull, 55

Rainbow lorikeet, its movements, 76; in captivity, 76–7

Red, its significance in birds' plumage, 24

Red-capped robin, nesting in dry weather, 50–1; its beauty, 51; its airy note, 51–2; breeding in immature plumage, 52; its fiery nature, 52; its comings and goings, 115–16

Red-backed parrot, 167

Red-winged parrot, confusion of names, 159

Reed-warbler, as a spring visitor, 22; singing at night, 24; its voice interpreted, 25

Regent honeyeater, nesting in September, 26; its beauty of color and voice, 27; community of

interest, 28; its capricious movements, 77
Restless flycatcher, its call and manner, 5–6; its feeble nest-building, 150
Robins, in winter, 4; their many varieties, 105; value of the name, 106–7; their curious distribution, 115; their nesting, 116; length of time in coloring, 116 (*See also* black-and-white, flame, pink, red-capped, rose, scarlet, and yellow robins.)
Rose-breasted robin, its movements, 117; its faint voice, 117; its winnowing of wings, 118; its butterfly flight, 118; its pugnacity, 119
Rosella parrots, nipping off wattle blossom, 14; their abundance, 167
Rufous-breasted whistler, as a spring visitor, 24, 32, 129; its distribution, 128–9; coloration of the sexes, 129; its ringing whistle, 129; female as songster, 129

Sanctuaries, birds' tameness therein, 154
Sarsaparilla, festooning trees, 13
Scaly-breasted lorikeet, 76
Scarlet-breasted robin, in verse, 52; its movements, 51, 115
Scarlet-chested parrot, 167
Scarlet honeyeater, its beauty of voice and plumage, 78; its love of the tropics, 78–9; behavior at nest, 79–80; feigning to be injured, 79
School children, contrasted, 59
School journals, of Australia, 60
School teachers, puzzled, 20; opinions of Nature-study, 58–9
Searches, for nests, 19; for shrike-tits' nests, 94; for the paradise parrot, 158–9

Seasons, their effect on the world, 43–4
Selection, sexual, 122
September, its festal opening, 17–8
Shrike-tit, *see* crested shrike-tit
Silvereye, among almond blossoms, 9–10
Skimmer, *see* wood-swallow
Skylark, English, in Australia, 23
Song-birds, in autumn, 22; their plumage discussed, 121–4
Sparrows, imposing on shrike-tits, 101
Spinebill, clipping wings, 14; its long bill, 14
Spring, manner of advent, 3; its impalpable heralds, 6; its sternness in 1914, 13, 33; its promise fulfilled, 32: its passing, 43; American and Australian contrasted, 31–2
Summer-bird, *see* wood-swallow
Sun-bird, its beauty and vivacity, 80; its fraternal nesting habits, 81
Swift lorikeet, 8

Tameness of birds, 61, 98, 108, 153–4
Tasmania, its robins, 115; lack of black-and-white birds, 141
Tea tree, as host for birds, 78
Tit-warblers, greeting the spring, 4; nesting, 14
Thickhead, *see* whistler
Thomas, James (poet), on the wagtail, 149
Thrush, English, in Australia, 45 (*See also* grey thrush.)
Trapping, effect on parrots, 157, 168
Travelled birds, as spring vocalists, 19, 24, 25
Tree-creepers, adaptability to dry

INDEX

conditions, 45; their pretty eggs, 45; faithfulness to nesting sites, 46; cleverness at nest, 46–7; fright of mother bird, 46

Tree-tit, pursued by robin, 119

Trumpet-bird, *see* manucode

Turquoisine parrot, its disappearance, 167; inquiry and search for it, 167–8

United States, of America, interest in Nature-study, 57

Victoria, its spring birds, 6

Voice, of birds, index to disposition, 8, 19, 24, 108, 123

Wagtail, its prominence and pluck, 137, 147; its alliance with the magpie-lark, 147–8; its nesting, 34–5, 150; in poetry, 149; its alliance with animals, 149; its calling at night, 149; its trip of 70 miles, 150

Warning, by birds, 29–30

Warning colors, Wallace theory of, 113

Wattle blossom, its beauty and fragrance, 13; harboring a birds' nest, 14

Wax-flower, 13

Wedgebill, 88

Whip-bird, 88–9

Whisperings, among robins, 15; among birds and flowers, 32

Whistlers, their numbers in Australia, 122; their plumage and voice, 123 (*See also* Gilbert, golden-breasted, and rufous-breasted whistlers.)

White, C. T., in search of a parrot, 159

White-fronted chat, greeting the spring, 4; feeding cuckoo, 54; its coloration, 140; feigning to be injured, 140

White-naped honeyeater, concert by, 19; its blithesome notes, 82

White-shouldered caterpillar-eater, as a spring visitor, 39; its cumbersome name, 67, 150–1; singing while flying, 39; nesting habits, 40; fondness for favorable situations, 41

Winter, its excursions, 17; its veil withdrawn, 32

"Winter's Tale," Shakespeare's, 108

Wiree, *see* rufous-breasted whistler

Witchery, of spring, 32

Wood-pecker, *see* tree-creeper

Wood-swallows, swarming like bees, 48; arrival in spring, 35–7; nesting habits, 38; nesting in hot weather, 51; as honeyeaters, 86

Year, interest in its changing moods, 55

Yellow, its light in dark places, 113–14

Yellow-robin, its busy days, 15; feigning to be injured, 28–9; dive from nest, 29; its pleasant scientific name, 107; its chanting at dawn, 108; its pretty housekeeping, 109–12; crouching on nest, 111; as host for a cuckoo, 112; pertinacity of a pair, 111–12; curious undulation in color, 113; clinging to a man's lips, 114–15; its clipping wings, 114

Yellow-tufted honeyeater, nesting, 14, 84–6; feigning to be injured, 28–9, 83; dancing, 82–3; the Autolycus of the bird-world, 85; as weather-prophet, 85; pursuing a shrike-tit, 101; pursuing a cuckoo, 8

LIST OF SCIENTIFIC NAMES

(In accord with the 1922 Check List of the Royal Australasian Ornithologists' Union.)

Babbler ("arcoe," "chatterer")	*Pomatorhinus temporalis*
Babbler ("cattie," "happy family")	*Pomatorhinus superciliosus*
Bee-eater (rainbow-bird)	*Merops ornatus*
Bell-bird, crested	*Oreoica cristata*
Bell-minah	*Manorhina melanophrys*
Blackbird, European (merle)	*Merula merula*
Bluebird, American	*Sialia sialis*
Bobolink, American	*Dolichonyx oryzivorous*
Bower-bird, satin	*Ptilonorhynchus holosericeus*
Bower-bird, spotted	*Chlamydera maculata*
Budgerigah (grass parrot)	*Melopsittacus undulatus*
Bush-lark (Australian skylark)	*Mirafra horsfieldi*
Butcher-bird, black-throated	*Cracticus nigrogularis*
Butcher-bird, grey ("whistling Jack")	*Cracticus destructor*
Caterpillar-eater, pied	*Campephega leucomela*
Caterpillar-eater, white-shouldered	*Campephega humeralis*
Chat, white-fronted	*Epthianura albifrons*
Chickadee, American	*Penthestes atricapillus*
Cockatoo, pink (cockalerina)	*Cacatua leadbeateri*
Cockatoo, sulphur-crested	*Cacatua galerita*
Cockatoo-parrot (cockatiel)	*Calopsitta novae-hollandiae*
Cuckoo, bronze	*Chalcococcyx plagosus*
Cuckoo, narrow-billed bronze	*Chalcococcyx basalis*
Cuckoo, fantailed	*Cacomantis flabelliformis*
Cuckoo, pallid	*Cuculus pallidus*
Cuckoo-shrike ("blue jay," "lapwing")	*Graucalus melanops*
Diamond-birds	*Pardalotus*
Dollar-bird (roller)	*Eurystomus pacificus*
Dottrel, black-fronted	*Aegialitis nigrifrons*
Dove, peaceful	*Geopelia tranquilla*
Fantail, black-and-white (wagtail)	*Rhipidura motacilloides*
Fantail, white-shafted ("cranky Fan")	*Rhipidura albiscapa*
Finch, spotted-sided ("fire-tail")	*Staganopleura guttata*
Flycatcher, brown (Jacky Winter)	*Micræca fascinans*
Flycatcher, restless ("grinder")	*Seisura inquieta*